AF845391

Wave Propagation in Materials and Structures

Wave Propagation in Materials and Structures

Editor

Emeka Arendse

scitus
academics

Wave Propagation in Materials and Structures
Edited by **Emeka Arendse**

ISBN: 978-1-68117-712-0
Library of Congress Control Number: 2016958825

© 2018 by
SCITUS Academics LLC,
616, Corporate Way, Suite 2, 4766,
Valley Cottage, NY 10989

www.scitusacademics.com

This book contains information obtained from highly regarded resources. Copyright for individual articles remains with the authors as indicated. All chapters are distributed under the terms of the Creative Commons Attribution License, which permits unrestricted use, distribution, and reproduction in any medium, provided the original author and source are credited.

Notice

Reasonable efforts have been made to publish reliable data and views articulated in the chapters are those of the individual contributors, and not necessarily those of the editors or publishers. Editors or publishers are not responsible for the accuracy of the information in the published chapters or consequences of their use. The publisher believes no responsibility for any damage or grievance to the persons or property arising out of the use of any materials, instructions, methods or thoughts in the book. The editors and the publisher have attempted to trace the copyright holders of all material reproduced in this publication and apologize to copyright holders if permission has not been obtained. If any copyright holder has not been acknowledged, please write to us so we may rectify.

Printed in United States of America on Acid Free Paper

Preface

The propagation of waves is one of the ways waves travel. As regards the path of oscillation in relation to the direction of propagation, we can distinguish between longitudinal wave and transverse wave. Propagation in a vacuum and in a material medium can occur in electromagnetic waves. Other types of waves cannot propagate through a vacuum and you must have a transmission medium. There has been an increasing interest in micro-and nanotechnology in recent decades. Nanotechnology offers advances in new types and new applications of materials with unique electromagnetic and physical characteristics. Structural and material wave propagation is dedicated to modern techniques developed to describe wave propagation in modern materials and structures. Original works by prominent wave propagation scientists who have developed new theoretical and experimental research techniques and achieved important and new results are included. The propagation of elastic waves is still used to assess the structural integrity of engineering materials for years. The advantage it offers is the direct association with elastic properties, the relatively easy application through commercial equipment together with numerous empirical correlations between pulse velocity and material strength or possibly common quality. Advanced features as frequency dependence on wave parameters can further increase the capacity of the characterization. Concrete materials improved by the existence of damage-induced cracking due to their intrinsic microstructure demonstrate a complex behavior regarding the propagation of pulses of different frequencies. The different wave lengths interact with each other in homogeneities depending on their size, leaving their signature on phase velocity and attenuation versus frequency curves. Although experimental measurements are difficult in concrete structures, mainly due to the attenuation of higher frequencies, it is recommended that the application of different frequencies can provide a much more detailed insight into the internal state of the structure whenever possible.

This book aims at advanced and basic wave propagation concepts in various material systems and structures. Such a text is necessary in view of the multidisciplinary nature of the subject. Many interesting results and new ideas in this book are therefore intended for readers (scientists, students, engineers, as well as researchers and practitioners) who are active.

Table of Contents

1

CHAPTER

Wave Propagation in Nanocomposite Materials

Pierre Hillion

Institut Henri Poincaré, Bis Route de Croissy, Le Vésinet, France.

ABSTRACT

Electromagnetic wave propagation is first analyzed in a composite material mde of chiral nano-inclusions embedded in a dielectric, with the help of Maxwell-Garnett formula for permittivity and permeability and its reciprocal for chirality. Then, this composite material appears as an homo-geneous isotropic chiral medium which may be described by the Post constitutive relations. We analyze the propagation of an harmonic plane wave in such a medium and we show that two different modes can propagate. We also discuss harmonic plane wave scattering on a semi-infinite chiral composite medium. Then, still in the frame of Maxwell-Garnett theory, the propagation of TE and TM fields is investigated in a periodic material made of nano dots immersed in a dielectric. The periodic fields are solutions of a Mathieu equation and such a material behaves as a diffraction grating.

Keywords: Composite materials, Maxwell-Garnett, Constitutive relations, TE TM Fields

1. Introduction

Nanotechnology is blossoming with in particular the inclusion of nano-particles (nano-dots) in some specific support [1,2]. Then, to analyze electromagnetic wave propagation such as light or X rays in these composite materials, we need a theory able to calculate average, macroscopic values from their granular microscopic properties. This job is performed by the Maxwell-Garett theory [3-6].

In this work, the support is a dielectric with permittivity e_1, permeability m and we consider two situations according that nano-dots are chiral or periodically distributed along a direction of the structure.

In the first case (chiral nano-dots) the permittivity e of the composite material is according to the MaxwellGarett formula

$$[e - e_1][e + 2e_1]^{-1} = f\,[e_2 - e_1][e_2 + 2e_1]^{-1} \qquad (1)$$

f is the filling factor of inclusions (their volume fraction) in the host material, the subscripts 1, 2 corresponding to host and inclusions respectively and we get from (1)

$$e = e_1\,(1 + 2af\,)(1 - af)^{-1}a = (e_1 - e_2)(e_1 + 2e_2)^{-1}, \qquad (2)$$

Permeability μ is assumed the same for nano-dots and dielectric.

The relation (2) has been generalized [7,8] to chirality x when both inclusions and host materials are chiral. But here, the situation is different since only inclusions have this property and the relation (1) with x, x_1, x_2 instead of e, e_1, e_2 has no meaning when $x_1 = 0$. To cope with this difficulty, we introduce a reciprocal Maxwell-Garnett relation obtained by applying to (1) the transformation (x_1, x_2, f) Þ $(fx_2, x_1, 1/f)$ which gives

$$[x - fx_2][x + 2fx_2]^{-1} = f^{-1}[x_1 - fx_2][x_1 + 2fx_2]^{-1} \qquad (3)$$

reducing for

$$x_1 = 0 \text{ to } x = -2fx_2(1 - f)(1 + 2f)^{-1} \qquad (4)$$

From now on, we assume $f << 1$, $e_1 > 0$, $e_2 < 0$, $a > 0$ and $x_2 < 0$ so that the $0(f^2)$ approximation of (2) and (4) gives $e = e_1(1 + 3af) > 0$(5)

$$x = -2fx_2 > 0 \tag{6}$$

So, this composite material made of nano-chiral particles included in a dielectic may be hand-led as an homogeneous chiral medium with permittivity and chirality (5) and (6) and permeabiity $\mu > 0$ assumed to be the same for inclusions and dielectric.

In the second situation (periodically distributed nanorods), the relation (1) is still valid with f changed into a periodic function f(x). Assuming $f(x) = f\cos(2ax)$, we write the permittivity e(x) in the following form reducing to (5) to the $0(f^2)$ order $e(x) = e_1\exp[3af\cos(2ax)]$(7)

Using (5)-(7), we shall analyze harmonic plane wave propagation in both composite materials, chiral and periodic.

2. Harmonic Plane Wave Propagation in a Chiral Composite Medium

We suppose this chiral medium endowed with the Post constitutive relations in which e, x have the expressions (5) and (6) [9,10]

$$D = eE + ixB,\ H = B/\mu + ixE,\ i = \sqrt{-1} \tag{8}$$

This choice is not arbitrary because the Post constitutive relations, in their general form, are covariant under the proper Lorentz group as Maxwell's equations which guarantees a consistent theory with a simple mathematical formalism, in agreement with the statement that only covariant mathematical expressions have a physical meaning.

Plane wave scattering from a semi-infinite chiral medium was discussed some time ago by Bassiri et al [11], also using the Post constitutive relations, but we proceed differently from these authors working with the Fresnel reflection and transmission amplitudes.

2.1 Refractive Indices

We consider harmonic plane waves with amplitudes E, B, D, H

$$(E,B,D,H)(x,t) = (E,B,D,H)y(x,t) \tag{9}$$

in which

$$y\,(x,t) = \exp[iw(t + n\sin\theta\, x/c + n\cos\theta\, z/c)] \tag{10}$$

in which n is a refractive index to be determined.

Substituting (9) into the Maxwell equations

$$\text{ÑÙ}E + 1/c\theta_t B = 0,\ \text{Ñ}.B = 0$$

$$\text{ÑÙ}H - 1/c\theta_t D = 0,\ \text{Ñ}.D = 0 \tag{11}$$

and taking into account (10) give the equations for the amplitudes E, B, D, H

$$-n\cos\theta\, E_y + B_x = 0,$$

$$n(\cos\theta\, E_x - \sin\theta\, E_z) + B_y = 0,$$

$$n\sin\theta\, E_y + B_z = 0, \tag{12}$$

$$n\cos\theta\, H_y + D_x = 0$$

$$n(\cos\theta\, H_x - \sin\theta\, H_z) - D_y = 0$$

$$n\sin\theta\, H_y - D_z = 0 \tag{13}$$

with the divergence equations

$$\sin q\, B_x + \cos q\, B_z = 0,\ \sin q\, D_x + \cos q\, D_z = 0 \tag{14}$$

We get at once from (8) and (14), the divergence equation satisfied by the electric field

$$\sin q\, E_x + \cos q\, E_z = 0 \tag{15}$$

Substituting (8) into (13) gives

$n\cos q\ (B_y/m + ixE_y) + eE_x + ix\ B_x = 0\ n\cos q\ (B_x/m + ixE_x) - n\sin q\ (B_z/m + ixE_z) - eE_y - ixB_y = 0\ n\sin q\ (B_y/m + ixE_y) - eE_z - ixB_z = 0$(16)

Taking into account (12), these equations become

$$n\cos q\ B_y/m + eE_x + 2ixB_x = 0\ n\cos q\ B_x/m - n\sin q\ B_z/m - eE_y - 2ixB_y = 0\ n\sin q\ B_y/m - eE_z - 2ixB_z = 0 \quad (17)$$

Then, eliminating B between (12) and (17) gives the homogeneous system of equations in which $s = 2nx$

$$(n^2/m - e)E_x - is\cos q\ E_y = 0$$

$$(n^2/m - e)E_y - is(\sin q\ E_z - \cos q\ E_x) = 0$$

$$(n^2/m - e)E_z + is\sin q\ E_y = 0 \quad (18)$$

This homogeneous system has nontrivial solutions if its determinant is null and a simple cal-culation gives

$$(n^2/m - e)[(n^2/m - e)^2 - s^2] = 0 \quad (19)$$

Deleting $(n^2/m - e) = 0$ which would correspond to an a-chiral medium, we get from (11) two modes $(n_\pm^2/m - e) = \pm s$ in which which $s = 2nx$ so that the refractive index depends not only on permittivity and permeability but also on chirality with the positive expressions

$$n_+ = xm + (x^2m^2 + em)^{1/2}, \quad (20)$$

$$n_- = -xm + (x^2m^2 + em)^{1/2} \quad (21)$$

Changing the square root into its opposite gives negative refractive indices.

Consequently, two modes with respectively the refractive indices n_+, n_- can propagate in the metachiral slab, they are independent as long as the medium is infinite, otherwise they become coupled at boundaries. The amplitudes of the field components in these two modes have now to be determined.

2.2 Electromagnetic Fields

1) We first suppose $n_+^2/m - e = s$ and $n_+ = xm + (x^2m^2 + em)^{1/2}$ with e and $\mu > 0$: fields and parameters are characterized by superscripts or subscripts + respectively.

Then, we get at once from (18) and (12) in terms of E^+_y

$E^+_x = i\cos q_+ E^+_y$, $E^+_z = - i\sin q_+ E^+_y B^+_x = n_+ \cos q_+ E^+_y B^+_y = -i\, n_+ E^+_y$, $B^+_z = - n_+ \sin_+ E^+_y$ (22)

and substituting (22) into (8)

$$D^+_x = i\cos q_+ l_+ E^+_y D^+_y = l_+ E^+_y D^+_z = -i\sin q_+ l_+ E^+_y \quad (23)$$

$$H^+_x = \cos q_+ n_+ E^+_y H^+_y = -i\, n_+ E^+_y H^+_z = -\sin q_+ n_+ E^+_y \quad (24)$$

$$\text{in which } l_+ = e + xn_+,\ n_+ = n_+/m - x = (x^2 + e/m)^{1/2} \quad (25)$$

2) For $n_-^2/m - e = - s$ and $n_- = - xm + (x^2m^2 + em)^{1/2}$, we get at once with now super-scripts and subscripts -:

$E^-_x = - i \cos q_- E_y$, $E^-_z = i \sin q_- E^-_y B^-_x = n_- \cos q_- E^-_y$, $B^-_y = i\, n_- E^-_y B^-_z = - n_- \sin q_- E^-_y$ (26)

and substituting (26) into (8)

$$D^-_x = -i\cos q_- l_- E^-_y,\ D^-_y = l_- E^-_y,$$

$$D^-_z = i\sin q_- l_- E^-_y (27)$$

$$H^-_x = \cos q_- n_- E^-_y,\ H^-_y = i\, n_- E^-_y,$$

$$H^-_z = - \sin q_- n_- E^-_y \quad (28)$$

$$\text{with } l_- = e - xn_-,\ n_- = n_-/m + x = (x^2 + e/m)^{1/2} = n_+ \quad (29)$$

Then, according to (9) and (10), the electromagnetic field of the plus and minus modes, each depending on an arbitrary amplitude E^+_y, E^-_y, is

$$(E^{\pm}, B^{\pm}, D^{\pm}, H^{\pm})\ (x,t) = (E^{\pm}, B^{\pm}, D^{\pm}, H^{\pm})\ y_{\pm}(x, t) \qquad (30)$$

with the amplitudes given by (22)-(24) and (26)-(28) and the phase functions

$$y_{\pm}(x,t) = \exp[iw\,(t + n_{\pm}\, \sin q_{\pm}\, x/c + n_{\pm}\, \cos q_{\pm}\, z/c)] \qquad (31)$$

2.3 Plane Wave Scattering from a Semi-Infinite Chiral Composite Medium

We suppose that the chiral composite material fulfills the half space $z < 0$ on which impinges from $z > 0$ on the interface $z = 0$ an harmonic plane wave characterized by the phase factor $y(q_i)$

$$y\,(q_i) = \exp[-iwn_0(x\sin q_i + z\cos q_i)] \qquad (32)$$

n_0 is the refractive index in $z > 0$ and the components of the incident electromagnetic field are [12] with two amplitudes M_i, N_i:

$$E^i{}_x = -\cos q_i\ M_i\, y\,(q_i),\ E^i{}_y = N_i\, y\,(q_i),\ E^i{}_z = \sin q_i\ M_i\, y\,(q_i)$$

$$H^i{}_x = -n_0\cos q_i\ N_i\, y\,(q_i),\ H^i{}_y = -n_0 M_i\, y\,(q_i) H^i{}_z = n_0 \sin q_i\ N_i\, y(q_i) \qquad (33)$$

The reflected field in the half-space $z > 0$ has a similar expression with (M_i, N_i, q_i) changed into (M_r, N_r, q_r) while the refracted field in $z < 0$ is supplied by (30).

According to (31) and (32), also valid for the reflected wave, the continuity of the phase at $z = 0$ implies the Descartes-Snell relations

$$n_0\ \sin q_i = n_0\ \sin q_r = n_+\ \sin q_+ = n_-\ \sin q_- \qquad (34)$$

The continuity of the components $E_{x,y}$, $H_{x,y}$, at $z = 0$ supplies four boundary conditions to de-termine in terms of M_i, N_i the amplitudes M_r, N_r of the reflected field and those $E^+{}_y$, $E^-{}_y$ of the refracted field.

According to (22), (26) and (33) and taking into account (34), we get for the $E_{x,y}$ components $\cos q_i(M_r - M_i) = i\cos q_+\, E^+{}_y - i\cos q_-\, E^-{}_y$

$$N_r + N_i = E^+{}_y + E^-{}_y \qquad (35)$$

while for $H_{x,y}$, according to (24), (28) and (33), we have since $n_- = n_+$ (= n)

$$n_0 \cos q_i(N_r - N_i) = n (\cos q_+ E^+{}_y + i\cos q_- E^-{}_y)$$

$$n_0(M_r + M_i) = n(E^+{}_y - E^-{}_y) \qquad (36)$$

To make calculations easier, we introduce the notations

$$M_r + M_i = M, N_r + N_i = NM_r - M_i = M', N_r - N_i = N' \qquad (37)$$

and

$$a = n_0/n (\cos q_+ + \cos q_-)^{-1} \qquad (38)$$

Then, we get at once from (36)

$$E^+{}_y = a(\cos q_i N' + \cos q_- M)$$

$$E^-{}_y = a(\cos q_i N' - \cos q_+ M) \qquad (39)$$

and, substituting (39) into (35) gives

$$\cos q_i M' = a_{11} N' + a_{12} M N = a_{21} N' + a_{22} M \qquad (40)$$

in which $a_{11} = ia\cos q_i (\cos q_+ + \cos q_-)$, $a_{12} = 2ia\cos q_+ \cos q_-$

$$a_{21} = a\cos q_i, a_{22} = a(\cos q_- + \cos q_+) \qquad (41)$$

Taking into account (37) the system (40) becomes

$$(\cos q_i - a_{12})M_r + a_{11} N_r = (\cos q_i + a_{12}) M_i - a_{11} N_i$$

$$a_{22}M_r - (1 - a_{21}) N_r = -a_{22}M_i + (1 + a_{21})N_i \qquad (42)$$

from which we easily get the amplitudes M_r, N_r of the reflected field and consequently M', N' according to (37) to obtain finally the amplitudes $E_y^{\pm}$ of the refracted field from (39).

One has a simple result for a normal incidence $q_i = q_r = q_\pm = 0$ since the Equations (35) and (36) reduce to $M_r - M_i = i(E^+_y - E^-_y)$, $N_r + N_i = E^+_y + E^-_y$

$$-2n_0 N_i = n(E^+_y + E^-_y),\ n_0(M_r + M_i) = n(E^+_y - E^-_y) \qquad (43)$$

with the solution

$$M_r = (n + in_0)(n - in_0)^{-1} M_i,\ N_r = -(1 + 2n_0/n)N_i \qquad (44)$$

$$E^+_y = n_0(n - in_0)M_i - n_0/n\ N_i,$$

$$E^-_y = -n_0(n - in_0)M_i - n_0/n\ N_i \qquad (45)$$

Remark 1. If the angles q_+, q_- obtained from (34) are real, the plus and minus modes propa-gate in the chiral medium. If they are both purely imaginary, we get from (34)

$$\cos(q_\pm) = -i[(n_0/n_\pm)^2 \sin^2 q_i - 1]^{1/2} \qquad (46)$$

the negative sign in front of the square root in (46) corresponds to the physical situation: refracted waves are evanescent and, incident waves undergo a total reflection, with as consequence for beams of plane waves a GoösHanken lateral shift and a Imbert-Fedorov transverse shift [13]. Of course with a single angle pure imaginary, only one mode propagates, the other mode giving rise to an evanescent wave.

Remark 2. At the expense of more intricacy, the present formalism may be generalized to wave propagation in a chiral slab located between z = 0 and z = - d. Then, two more fields exist respectively reflected at z = -d inside the slab and refracted outside in the z < -d region, supplying four supplementary amplitudes matched by the boundary conditions at z = -d. But, instead of a 4 ´ 4 system of equations to get the amplitudes of the electromagnetic field, we have to deal with a 8 ´ 8 system more difficult to solve.

3. Harmonic Plane Wave Propagation in a Two Dimensional Nano-Periodic Medium

With B = μH, D = e(x)E, and exp(-iwt) implicit, the Maxwell equations are for E(x,z), H(x,z)

$$\theta_z E_y - iwm/c\ H_x = 0,\ \theta_z H_y + iw\ e(x)/c\ E_x = 0$$

$$\theta_z E_x - \theta_x E_z + iwm/c\ H_y = 0 \theta_z H_x - \theta_x H_z - iw\ e(x)/c\ E_y = 0$$

$$\theta_x E_y + iwm/c\ H_z = 0,\ \theta_x H_y - iwe(x)/c\ E_z = 0 \quad (47)$$

with the divergence equations

$$[e' + e\theta_x]E_x + e\theta_z E_z(x,z) = 0,\ \theta_x H_x + \theta_z H_z = 0 \quad (48)$$

giving rise to TE (E_y, H_x, H_z) and TM (H_y, E_x, E_z) waves.

3.1 TE Wave Propagation

Assuming f << 1, we work with the Maxwell-Garnett $0(f^2)$ approximation of (7)

$$e(x) = e_1 + h\ f \cos(2ax),\ h = 3ae_1 \quad (49)$$

The component E_y satisfies the Helmholtz equation in which $\Delta = \theta_x^2 + \theta_z^2$

$$[\Delta + w^2 me(x)/c^2]E_y(x,z) = 0 \quad (50)$$

We look for the solutions of this equation in the form, A being an arbitrary amplitude

$$E_y(x,z) = A \exp(ik_z z)\ y(x) \quad (51)$$

Substituting (51) into (50) and taking into account (49), gives the differential equation satisfied by y(x)

$$[\theta_x^2 + k_0^2 + f\ k_e^2 \cos(2ax)]\ y(x) = 0 \quad (52)$$

in which

$$k_0^2 = w^2 me_1/c^2 - k_z^2,\ k_e^2 = w^2 mh/c^2 \qquad (53)$$

Using the variable $z = k_1 x$, Equation (52) becomes a Mathieu equation [14,15]

$$[\theta_z^2 + c^2 + f\cos(2az/k_e)]y(z) = 0,\ c^2 = k_0^2/k_e^2 \qquad (54)$$

with solutions in the form [14,15,16] where v has to be determined y

$$(z) = \sum_{m=-¥}^{¥} c_m \exp([i(v + 2m)\,az/k_e)] \qquad (55)$$

Substituting (55) into (54) gives the following recurrence relation [15] for the coefficients c_m

$$c_m + g_m(v)\,(c_{m-1} + c_{m+1}) = 0 \qquad (56)$$

$$\text{with } g_m(v) = -f/2\,[(2m + v)^2 - c^2] \qquad (57)$$

Now, the main difficulty [14,15] is to get v in terms of f and c, but f being small, the infinite determinant of the system (56) supplies v to the $0(f^3)$ order [15]

$$\cos(v\pi) = \cos(c\pi) + \pi f^2\,[4c^2(1 - c^2)^{1/2}]^{-1}\sin(c\pi) \qquad (58)$$

Once v known, the c_m coefficients may be obtained by numerical methods based on the recurrence relations (36) or on some variant of it. It is shown [15] how for moderate values of c and f, these relations can be transformed into convergent continued fractions $R_m(n) = c_m/c_{m-1}$, $L_m(n) = c_m/c_{m+1}$.

So, according to (51) and (55), $E_y(x,z) = E_y(x + \pi/a, z)$ and $E_y(x,z) = A \exp(ik_z z) \sum_{m=-¥}^{¥} c_m \exp[i(v + 2m)ax)] 0 \le x < \pi/a$ (59)

and taking into account the Maxwell Equation (47), the other two components H_x, H_z of the TE field are obtained from $\theta_z E_y$ and $\theta_x E_y$ respectively. Writing (59)

$$E_y(x,z) = A\sum_{m=-¥}^{¥} c_m \exp(ik_z z + ik_m x) k_m = (v + 2m)a,\ 0 \le x < \pi/a \qquad (60)$$

$E_y(x,z)$ appears as a periodic beam of plane waves propagating in the directions defined by the wave vectors with components (k_z, k_m), their amplitude being weighted by the coefficients c_m.

3.2 TM Wave Propagation

For TM waves (H_y, E_x, E_z), we start with the expression (7) of e (x). Then, according to the Maxwell Equation (47) the component H_y satisfied the equation

$$[\Delta + w^2 m e(x)/c^2 - \{e'(x)/e(x)\}\theta_x] H_y(x,z) = 0 \tag{61}$$

We look for the solutions of (61) in the form

$$H_y(x,z) = A \exp(ik_z z)\, y(x) \tag{62}$$

$$y(x) = u(x)\, f(x),\ f(x) = \exp[\, f_1/2 \cos(2ax)] f_1 = f\, h \tag{63}$$

A simple calculation gives the first and second derivative of y (x)

$$y'(x) = [u'/u - af_1 \sin(2ax)]y(x)$$

$$y''(x) = [u''/u - 2a\, u'/u\, f_1 \sin(2ax)$$

$$- 2a^2 f_1 \cos(2ax) + a^2 f_1^2 \sin^2(2ax)]y(x) \tag{64}$$

and since $e'/e = -2a\, f_1 \sin(2ax)$, we get to the $0(f_1^2)$ order $y'' - e'/e\, y' = [u'/u - 2a^2 f_1 \cos(2ax)]\, y(x) + 0(f_1^2)$ (65)

so that

$$[y'' - e'/e\, y']H_y(x,z) = [u'/u - 2a^2 f_1 \cos(2ax)]H_y(x,z) \tag{66}$$

Then, according to (62) and (66), we get from (61), the differential equation satisfied by u(x)

$$[\theta_x^2 + w^2 m e(x)/c^2 - k_z^2 - 2a^2 f_1 \cos(2ax)]u(x) = 0 \tag{67}$$

which becomes with the Maxwell-Garnett approximation (49) of e (x)

$$[\theta_x^2 + k_0^2 + f\, k_h^2 \cos(2ax)]u(x) = 0 \qquad (68)$$

with k_0^2 given by (53) while

$$k_h^2 = w^2 m e_1/c^2 - 2a^2 h \qquad (69)$$

The comparison of (52) and (68) shows that, to the $0(f^2)$ order, one has just to change k_e into k_h to go from TE to TM waves so that all the calculations of Subsection 3.1 can be repeated mutatis mutandis.

3.3 TE Wave Scattering in a Semi Infinite Nano-Periodic Material

The granular material, made of nano dots immersed in a dielectric, lies in the z < 0 half-space and we suppose that a TE harmonic plane wave (E_y^i, H_x^i, H_z^i) impinges from the upper half-space z > 0 with refractive index n and permeability m on the z = 0 interface.

The components E_y^i, E_y^r of the incident and reflected waves are

$$E_y^i(x,z) = A_i \exp[iwn/c\,(x \sin q_i + z \cos q_i)]$$

$$E_y^r(x,z) = A_r \exp[iwn/c\,(x \sin q_i - z \cos q_i)] \qquad (70)$$

and according to the Maxwell Equation (47), the components H_x^i, H_x^r involved in the boundary conditions are

$$H_x^i(x,z) = n/m \cos q_i\, E_y^i(x,z) H_x^r(x,z) = -n/m \cos q_i\, E_y^r(x,z) \qquad (71)$$

Now, the refracted periodic field in z < 0 has the form (59)

$$E_y^t(x,z) = A_t \exp(ik_z z) \sum_{m=-¥}^{¥} c_m \exp[i(v + 2m)\, ax] 0 \le x < \pi/a \qquad (72)$$

and, still using (47)

$$H_x^t(x,z) = g\, E_y^t(x,z),\ g = ck_z/wm \qquad (73)$$

the boundary conditions impose the continuity on z = 0 of E_y and H_x, that is, according to (70)-(73)

$$(A_i + A_r) \exp(iwn /c\ x\ \sin q_i)$$

$$= A_t \sum_{m=-¥}^{¥} c_m \exp[i(v + 2m)ax] 0 \leq x < \pi/a \quad (74)$$

$$n /m \cos q_i (A_i - A_r) \exp(iwn /c\ x\ \sin q_i)$$

$$= gA_t \sum_{m=-¥}^{¥} c_m \exp[i(v + 2m)ax] 0 \leq x < \pi/a \quad (75)$$

Let us write (74)

$$A_i + A_r = A_t W(b\ p) W(b\ p) = \sum_{m=-¥}^{¥} c_m \exp[i(k_m - k_i) ß\pi] \quad (76)$$

in which according to (60) and (70)

$$k_m = (v + 2m)a,\ k_i = wn /c \sin q_i \quad (77)$$

with $0 \leq ß < 1$ since $0 \leq x \leq \pi/a$.

So, taking into account (60), the granular periodic semi-infinite material behaves as a diffraction grating: the beam of plane waves propagating in the directions defined by the wave vectors with components (k_z, k_m) have their amplitudes modulated by the coefficients $c_m \exp(-ik_i ß\pi)$. And, acccording to (76), the relations (74) and (75) become $A_i + A_r = A_t W(b\ p)$, $n/m \cos q_i (A_i - A_r) = gA_t W(b\ p) 0 \leq ß < 1$(78)

from which we get in terms of the incident amplitude A_i

$A_r = -(gm - n \cos q_i)(gm + n \cos q_i)^{-1} A_i A_t = 2n \cos q_i (gm + n \cos q_i)^{-1} W^{-1}(b\ p) A_i$(79)

So, the amplitude A_t is not constant on the interval $(0, \pi/a)$.

4. Discussion

The relation (6), leads to a consistent formalism but further work is needed to prove or to amend it. In any case, two different modes of harmonic plane waves propagate in these chiral materials. The Post constitutive relations used to characterize such media, allow to get exact analytic expressions for the amplitude of the electromagnetic field in each mode, a note-worthy property due, as noticed in the introduction, to the covariance of Post's relations under the proper Lorentz group. An excellent review of chiral nano-technology may be found in [17] with a discussion of two topics: nanoscale approaches to chiral technology and, corresponding to the situation considered here, nanotechnology that benefits from chirality. In particular, a section is devoted to chiral carbon nanotechnology and the authors conclude "possible applications of such materials in the field of biomedecine and biotechnology range from prepara-tion of novel antibacterial, cyclotonic and drug delivery agents to catalysis and materials science applications".

Remark: The analysis of Section 2 may be performed in left-handed chiral materials with negative e, m: just change e, m into -|e|, |-m|.

Granular periodic materials are currently used in mechanical engineering and, with the ob-jective to appraise their properties, theoretical studies have been devoted to acoustic wave propagation in these structures [18]. In electrical engineering, photonic crystals [19,20] are the main illustration of periodic nanomaterials and they take an increasing importance in today technology. But, they are not composite with inclusions immersed in a dielectric structure. For instance, a one-dimensional photonic crystal with a permittivity periodic in the direction of propagation may be described by an expansion in which U is the unit step function $e(z) = e_1 \sum_n [U(z - 2na) - U(z - \{2n + 1\}a)]$

$$+ e_2 \sum_n [U(z - \{2n + 1\}a) - U(z - \{2n + 2\}a)] \qquad (80)$$

and, the solutions of Maxwell's equations are the Bloch functions $\sum_m c_{k,m} \exp(ikz+2i\pi mz/a)$ to be compared with (59) (and (80) with (49)). Incidently, (80) has a simple expression in terms of the square-sine function $e(z) = e + r \sin(az) / |\sin(az)|\ U(z) e_1 = e + r,\ e_2 = e - r$(81)

which suggests to work with the Laplace transform of Maxwell's equations since $\tanh(\pi p/2a)$ is the Laplace transform of the square-sine function [21]. People fluent with the Laplace transform, could think in terms of p instad of z as they use to do with w instead of t.

In opposite to photonic crystals, composite granular materials with a continuous filling factor have no lattice structure and, as shown in Section 3.3, they rather behave as a smooth dielectric grating [22]. Some of the restrictive assumptions on the filling factor f could be somewhat released at the expense of more intricacy:

1) It would be interesting to check what happens when a higher order approximation than $0(f^2)$ is used;

2) When $f(x) = f \cos(2ax)$ is changed into $f(x) = \sum_0^{¥} f_m(\cos 2max)$, the Mathieu equation becomes a Hill equation [14,15] with solutions similar to (55) but the recurrence relations bet-ween the coefficients c_m is more intricate;

3) Finally a generalization to a two-dimensional filling factor f(x,y), periodic in x and y would approach more closely a real physical situation.

To sum up, the application of the Maxwell-Garnett theory to nano composites deserves further research, taking into account the innocuity or not of such materials in biomedecine [23]. This theory is also used to analyze, in the frame of surface plasmon polaritons, the scattering of TE, TM light waves from a composite material made of metallic nano spherical particles immersed inside a metallic structure such as Ag particles in a Sio^2 matrix [24].

The $0(f^2)$ Maxwell-Garnett approximation of the periodic permittivity in the nanodoped medium of Section 3 implies that TE, and TM fields are solutions of the Mathieu equation as if they were diffracted from a dielectric grating [25].

REFERENCES

1. M. Khan, A. K. Sood, F. L. Deepak and C. N. R. Rao, "Nanorotors Using Asymmetric in Organic Nanorods in an Optical Trap," Nanotechnology, Vol. 17, No. 11, 2006, pp. S287-S290.

2. X. M. Guo, C. Jiang and T. S. Shi, "Prepared Chiral Nanorods of a Cobalt," Inorganic Chemistry, Vol. 46, No. 12, 2007, pp. 4766-4768.

3. J. C. Maxwell-Garnett, "Colours in Metal Glasses and Metallic Films," Philosophical Transactions of the Royal Society of London A, Vol. 203, No. 359-371, 1904, pp. 385-420.

4. C. A. Grimes, "Calculation of the Effective Electromagnetic properties of Granular Materials," in: A. Lakhtakia, Ed., Essays on the Formal Aspects of Electromagnetic Theory, World Scientific, Singapore, 1993, pp. 699-746.

5. L. Tsang and J. A. Kong, "Scattering of Electromagnetic Waves: Advanced Topics," Wiley Series in Remote Sensing, Wiley, New York, 2001.

6. P. Mallet, C. A. Guerin and A. Sentenac, "Maxwell-Garnett Mixing Rule in the Presence of Multiple Scattering," Physical Review B, Vol. 72, No. 1, 2005, p. 014205.

7. B. Shanker and A. Lakhtakia, "Extended Maxwell-Garnett Model for Chiral-in-Chiral Composites," Journal of Physics D: Applied Physics, Vol. 26, No. 10, 1993, pp. 1746- 1758.

8. A. Lakhtakia, V. K. Varadan and V. V. Varadan, "On the Maxwell-Garnett Model of Chiral Composite," Journal of Materials Research, Vol. 8, No. 4, 1993, pp. 917-922.

9. E. J. Post, "Formal Structure of Electromagnetics," NorthHolland Publications, Amsterdam, 1962.

10. A. Lakhtakia,V. K. Varadan and V. V. Varadan, "TimeHarmonic Electromagnetic Fields in Chiral Media," Springer, Berlin, 1989.

11. S. Bassiri, C. H. Papas and N. Engheta, "Electromagnetic Wave Propagation through a Dielectric-Chiral Interface and through a Chiral Slab," Journal of the Optical Society of America A, Vol. 5, No. 9, 1988, pp. 1450-1459.

12. M. Born and E. Wolf, "Principles of Optics," Pergamon Press, Oxford, 1965.

13. P. Hillion, "Light Beam Shifts in Total Reflection," Optics Communications,Vol. 266, No. 1, 2006, pp. 336-341.

14. E. T. Whittaker and G. N. Watson, "A Course of Modern Analysis," University Press, Cambridge, 1962.

15. A. Erdelyi, "Higher Transcendental Functions," McGrawHill, New York, Vol. 3, 1955.

16. P. Hillion, "Theoy of Electromagnetic Wave Propagation in a Longitudinally Periodic Cylinder," European Physical Journal B, Vol. 62, No. 4, 2008, pp. 477-480.

17. J. Zhang. M. T. Albelda, Y. Liu and J. W. Canary, "Chiral Nanotechnology," Chirality, Vol. 17, No. 7, 2005, pp. 404-420.

18. C. Inserra, V. Tournat and V. Gusev, "A Method of Controlling Wave Propagation in Initially Spatially Periodic Media," Europhysics Letters, Vol. 78, No. 4, 2007, p. 44001.

19. J. D. Joannopoulos, R. D. Meade and J. N. Wynn, "Photonic Crystals," University Press, Princeton, 1995.

20. C. Lopez, "Material Aspects of Photonic Crystals," Advanced Materials, Vol. 15, No. 20, 2003, pp. 1679-1704.

21. B. Van Der Pol and H. Bremmer, "Operational Calculus," Academic Press, Cambridge, 1959.

22. M. Neviere and E. Popov, "Light Propagation in Periodic Media," Marcel Dekker, Basel, 2005.

23. C. A. Poland, R. Duffin, I. Kinloch, A. Maynard, et al., "Carbon Nanotubes Introduced into the Abdominal Cavity of Mice Show Asbestos-Like Pathogenicity in a Pilot Study," Nature Nanotechnology, Vol. 3, 2008, pp. 423-428.

24. S. Kawata, "Near-Field Optics and Surface Plasmon Polariton," Springer, Berlin, 2001.

25. N. Yamaguchi, "High Order Bragg Diffraction by Dielectric Grating," Electronics and Communications in Japan, Vol. 69, No. 8, 2007, pp. 112-121.

2

CHAPTER

Propagation of Ultrasonic Guided Waves in Composite Multi-Wire Ropes

Renaldas Raisutis *, Rymantas Kazys, Liudas Mazeika, Vykintas Samaitis and Egidijus Zukauskas

Ultrasound Institute of Kaunas University of Technology, K. Barsausko 59, Kaunas Lt-51423, Lithuania

Abstract

Multi-wire ropes are widely used as load-carrying constructional elements in bridges, cranes, elevators, *etc*. Structural integrity of such ropes can be inspected by using non-destructive ultrasonic techniques. The objective of this work was to investigate propagation of ultrasonic guided waves (UGW) along composite multi-wire ropes in the cases of various types of acoustic contacts between neighboring wires and the plastic core. The modes of UGW propagating along the multi-wire ropes were identified using modelling, the dispersion curves were calculated using analytical and semi-analytical finite element (SAFE) techniques. In order to investigate the effects of UGW propagation, the two types of the acoustic contact between neighboring wires were simulated using the 3D finite element method (FE) as well. The key question of investigation was estimation of the actual boundary conditions between neighboring wires (solid or slip) and the real depth of penetration of UGW into the overall cross-section of the rope.

Therefore, in order to verify the results of FE modelling, the guided wave penetration into strands of multi-wire rope was investigated experimentally. The performed modelling and experimental investigation enabled us to select optimal parameters of UGW to be used for non-destructive testing.

Keywords: multi-wire ropes; modelling; ultrasonic guided waves; dispersion curves

PACS:

43.35.Yb; 43.35.Zc; 43.60.Qv

1. Introduction

Multi-wire ropes are widely used in different sectors of industry as load-carrying constructional elements in bridges, cranes, elevators, *etc.* Due to environmental and operational conditions and fatigue caused by rotational stresses, multiple defects may develop in the internal structure of multi-wire ropes. One of the promising methods for investigation of non-homogeneities of the internal structure of wire ropes can be application of ultrasonic guided waves (UGW) [1,2,3,4,5,6,7,8,9,10]. UGW are suitable for detection of various structural non-homogeneities, because they can propagate long distances and are quite sensitive both to surface and internal defects [11,12,13]. However, in the case of multi-wire ropes, the key issues are whether UGW will propagate only in one particular wire/strand, in which it had been excited, or in the whole composite multi-wire rope, and how the penetration of UGW will depend on the excitation method and the type of a mechanical contact between neighboring wires.

The objective of this work was to investigate the propagation of ultrasonic guided waves (UGW) along composite multi-wire ropes in the cases of various types of acoustic contacts between neighboring wires and the plastic core.

2. The Object of Investigation

A multi-wire steel rope to be investigated using ultrasound by itself is a complex composite structure in which multiple reflections, scattering, and the presence of multiple wave modes of UGW occur. Therefore, in order to understand the effects of UGW propagating along the multi-wire rope and penetration into the internal structure, determination of type of the acoustic contact between neighboring wires is necessary. In addition, the optimal parameters of UGW used for non-destructive testing of the particular objects are necessary to be investigated and selected.

As an object of investigation, a multi-wire steel rope consisting of six strands, each with the diameter of 11.15 mm, had been selected. The core of the multi-wire rope had been filled with polypropylene, and each strand consisted of a bundle of approximately 31 wires, each with the diameter of 2.05 mm. The cross-sectional view of the investigated composite multi-wire rope is presented in Figure 1. In order to simplify further analysis the assumption was made that the structure of rope consists of straight strands.

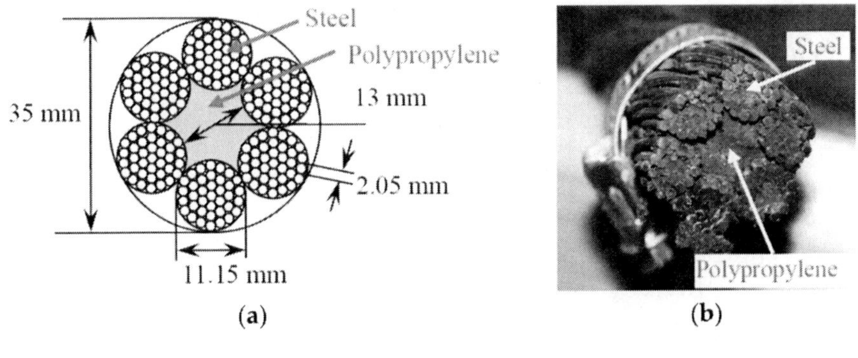

Figure 1. The rope under investigation: (**a**) schematic view; and (**b**) photo of the cross-section of the multi-bundle and multi-wire twisted rope.

2.1. Dispersion Curves of Propagating Guided Waves

Dispersion curves of propagating UGW along the object under investigation show the distribution of phase or group velocities in a particular frequency bandwidth. Therefore, in order to identify the particular modes of UGW, which propagate along the object under investigation and select suitable ones for non-destructive testing, the dispersion curves of phase and group velocities must be calculated. There are several techniques for calculation of the dispersion curves of UGW: semi-analytical finite element (SAFE) technique [14,15,16,17,18,19,20] and analytical ones [21]. The SAFE technique is more attractive for analysis of the objects with an arbitrary geometry of the cross-section (e.g., rails, bars, *etc.*). The SAFE technique has been used by Treyssede and Laguerre for investigation of UGW propagation along multi-wire helical strands with a metal cylindrical core [4,7,9,10,14]. The wires were assumed to be solidly connected [4,9,14]. They have found that the wave numbers and energy velocities of axisymmetric modes, such as $L(0,1)$ and $T(0,1)$, also flexural mode $F(1,1)$ at low frequencies and small helix lay angles (less than 15 degrees) of helical waveguides are very similar to the straight (not twisted) and twisted (helical) wires of multi-wire helical waveguides [4,9,14]. However, the cut-off region of the$L(0,1)$-like mode was observed at a notch frequency [4,9,14]. The advantage of the analytical technique, comparing it to SAFE, is a possibility to determine the frequency dependent guided wave leakage losses in the case of submerged wire ropes, for instance, as in cement grout or rock-embedded wire ropes [20,21].

In order to calculate dispersion curves for a multi-wire rope with a polymer core, several tasks are necessary to be solved. At first, the stranding of wires was eliminated and calculation of dispersion curves was divided into several stages. During the first stage, the dispersion curves of UGW were calculated using the SAFE technique for a single wire having the diameter of 2.05 mm (Figure 2a,b). The material properties used for the SAFE calculations of the dispersion curves are summarized in Table 1.

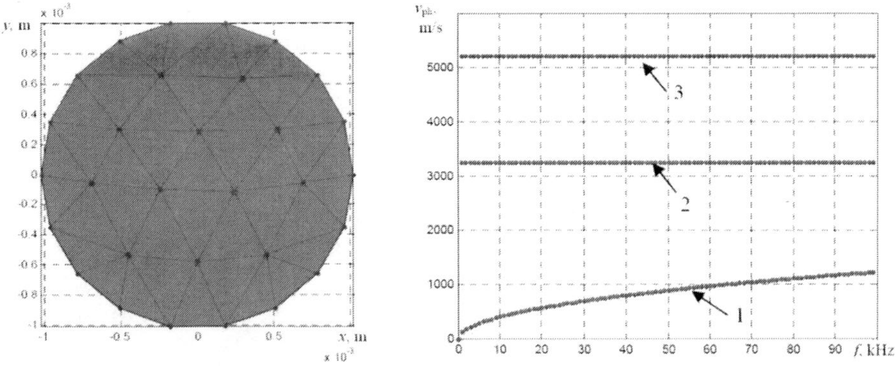

(a) (b)

Figure 2. The mesh of the SAFE model of a single wire with diameter of 2.05 mm (**a**) and the dispersion curves of the phase velocity of guided wave modes (**b**): 1—flexural $F(1,1)$ mode; 2—torsional $T(0,1)$ mode; 3—longitudinal $L(0,1)$ mode.

Table 1. Material properties between neighboring wires used for the calculation of the dispersion curves.

Material	Young's Modulus	Poisson's Ratio	Density
Steel	212.8 GPa	0.287	7800 kg/m^3
Polypropylene core	3.54 GPa	0.36	901 kg/m^3
"Elastic fluid"	0.075 MPa	0.499	1000 kg/m^3

During the SAFE calculations the triangular finite elements were used in order to obtain the desired accuracy, the average spatial size of element was set to 1 mm, which corresponds to 15 elements per wavelength, according to the minimum wavelength of the slowest $F(1,1)$ mode at 100 kHz frequency.

Calculation results of UGW propagation along a single wire having the diameter of 2.05 mm (Figure 2a) demonstrated that only fundamental modes propagate in the frequency band up to 100 kHz (Figure 2b). The dispersion curves also show that, in the selected frequency region, the fastest mode is longitudinal $L(0,1)$ and the slowest mode—flexural $F(1,1)$, whereas the torsional mode $T(0,1)$ is not dispersive.

The obtained results demonstrated that in this frequency range only fundamental modes propagate in a single wire. Typically, a multi-wire rope structure is more complicated, so, during the second stage the dispersion curves of the strand with 31 wires (the diameter of each wire 2.05 mm) having the overall diameter of 11.15 mm were calculated. In this case it was assumed that there is a solid acoustic contact between neighboring wires at their contact points. The cross-section of the meshed strand of 31 wires (the overall diameter 11.15 mm) is presented in Figure 3a. The obtained phase velocity dispersion curves and the comparison with the dispersion curves of the single wire (diameter 2.05 mm) are presented inFigure 3b. It should be noted that in the case of a bundle of wires or multi-wire rope the modes should be named as the axial-like $L(0,1)$, the flexural-like $F(1,1)$, and the torsional-like $T(0,1)$ modes.

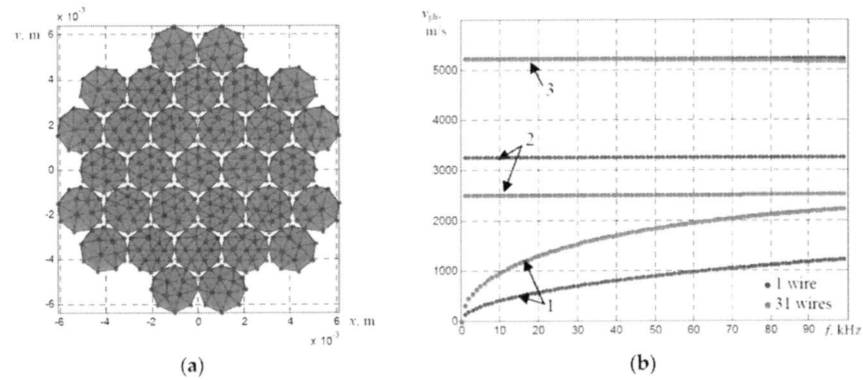

Figure 3. The mesh of the SAFE model of a bundle of solidly connected 31 wires (overall diameter 11.15 mm) (**a**) and comparison of the phase velocity dispersion curves of the guided wave modes (**b**) propagating along: one wire (**blue**), bundle of 31 wires (**red**): 1—flexural-like $F(1,1)$ mode; 2—torsional-like $T(0,1)$ mode; 3—axial-like $L(0,1)$ mode.

However, the numerical analysis of the multi-wire rope with the six strands consisting of 31 wires each and the polymer core is quite complicated due to limited computational resources of SAFE and the expected memory overflow. Therefore, the assumption was made that each strand can be replaced by a solid rod, having the same diameter (11.15 mm). The comparison of the dispersion curves of the UGW propagating in a multi-wire strand of 31 wires and in the solid rod are presented in Figure 4a,b. Comparison of these two cases demonstrate that the phase velocities of the $F(1,1)$ and $L(0,1)$ modes are similar and the velocity of the $T(0,1)$ mode in the strand of 31 wires is reduced by 23% in comparison to the phase velocities in the case of a solid rod. The calculation results demonstrate that a solid circular rod having the diameter of 11.15 mm is not fully equivalent to the meshed bundle of 31 wires but this replacement can be done for analysis of fundamental modes of UGW (especially $L(0,1)$ and $F(1,1)$) at frequencies below 100 kHz.

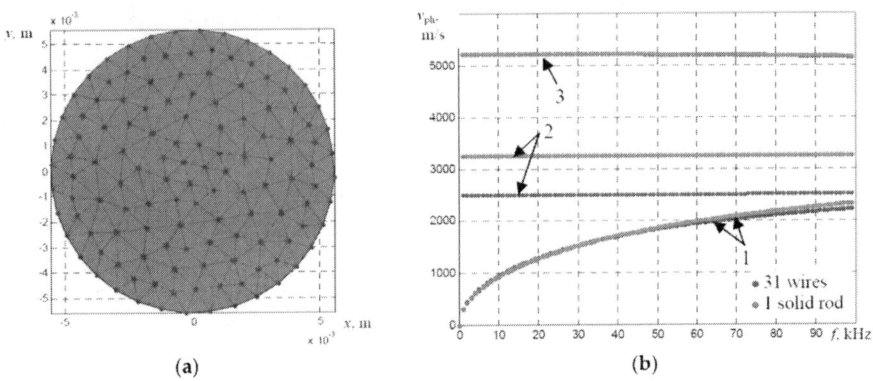

Figure 4. The mesh of the SAFE model of a strand that assumed to be equivalent to the bundle of solidly connected 31 wires having the overall diameter of 11.15 mm (**a**) and comparison of the phase velocity dispersion curves of the guided wave modes (**b**) propagating along the bundle of 31 wires (**blue**) and the solid rod of same diameter (**red**): 1—flexural-like $F(1,1)$ mode; 2—torsional-like$T(0,1)$ mode; 3—axial-like $L(0,1)$ mode.

This comparative analysis was made in order to be sure that replacement of the bundle of solidly-connected 31 wires by a single solid rod gives the same or similar results of the calculated dispersion curves. Then the dispersion curves of the whole multi-wire rope could be calculated as the curves of a bundle of solidly-connected six strands (the overall diameter 35 mm). This simplified assumption could reduce the required computational resources and additionally enable us to estimate frequency dependent leakage losses using the simplified analytical technique.

When conducting theoretical analysis on a multi-wire steel rope of a complex structure, an important question arises which is related to the acoustic contact between neighboring wires. Transmission of the appropriate modes between neighboring wires depends on a type of the acoustic contact. For example, in the case of a slip contact there is no transmission of the axial component of the particle velocity of the propagating mode to the neighboring wires. Therefore, only the component of the particle velocity perpendicular to the axis of the rope could be transmitted to the neighboring wires. In the case of a solid contact, the axial component is transmitted to neighboring wires. In order to investigate the influence of the acoustic contact on dispersion curves of the guided waves, the calculations for the simplified structure of the wire rope with the polymer core was performed. The calculations of the dispersion curves were performed for two separate cases. In Case A, it was assumed that the acoustic contact between neighboring wires is solid and in Case B a thin layer of a new material named as "elastic fluid" (Table 1) with the shear wave velocity close to zero was introduced between the separate strands to simulate the slip contact. The cross-section of the meshed multi-wire rope in the cases of a solid and slip contacts are presented in Figure 5a,b. The dispersion curves in the case of a solid and slip contact are presented in Figure 5c.

The results obtained demonstrate that the type of the contact between neighboring straight strands influences the type of modes that propagate in the structure. The phase velocity of the $L(0,1)$-like mode is the same for both cases at the frequencies below 30 kHz, whereas the phase velocity of the asymmetrical $F(1,1)$-like mode is shifted down in the case of the slip contact. It can be seen that in the case of the slip contact a large number of higher-order modes propagate in the structure at the frequencies above 10 kHz.

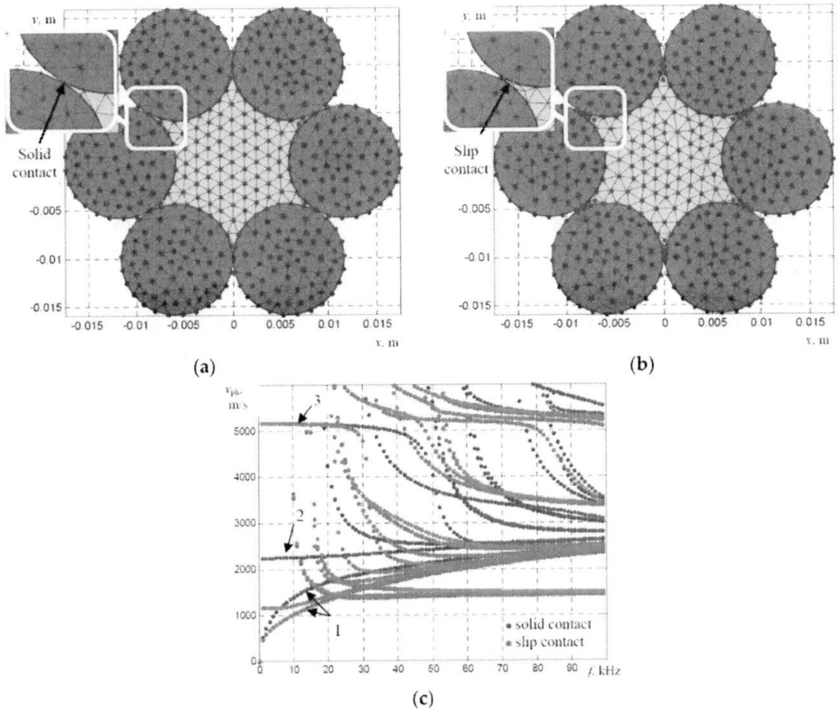

Figure 5. The mesh of the SAFE model of a multi-wire rope in the case of a solid (**a**) and a slip (**b**) contact between neighboring strands and comparison of the phase velocity dispersion curves of the guided wave modes (**c**) propagating along the particular structure: solid contact (**blue**), slip contact (**red**): 1—flexural-like $F(1,1)$ mode; 2—torsional-like $T(0,1)$ mode; 3—axial-like $L(0,1)$ mode.

2.2. Investigation of the Acoustic Contact Between Wires Using 3D FE Modelling

Dispersion curves of UGW show what modes may propagate along a multi-wire rope. The most appropriate way to excite UGW in such structures is to apply a normal (perpendicular to the axis of the rope) or tangential force to the outer surface of a wire rope (parallel to the axis of the rope). It is a promising solution for practical applications, based on a single side access to the surface of the rope. Thus, the next stage of investigation must be devoted to the issue whether those UGW modes can be excited from the outer surface and whether such propagating waves penetrate into the whole structure of a wire rope or just propagate only along the outer wires. In this case, a very important task is determination of the type of the acoustic contact between neighboring wires in a rope. Each wire is attached to another one by some force and in general, two types of acoustic contact between wires should be analyzed—a solid contact and a slip one. Those contact types influence what components (perpendicular to the axis of the rope or axial) of the particle velocity can propagate from one single wire to another one and, as a consequence, determine how deeply the propagating guided wave can penetrate into the structure under investigation. In order to get the answer to these questions, two models had been constructed using the finite element method with two different types of the acoustic contact between wires—a solid and a slip contact. In the case of the solid contact vertical (v_y) and axial (v_z) components of particle velocity are transmitted. Still, in the case of the slip acoustic contact only components (v_y) perpendicular to the rope axis are transmitted.

Elastic wave propagation can be described by the Navier equation of motion in the form of matrices [22]:

$$\mathbf{M}\ddot{\mathbf{U}} + \mathbf{C}\dot{\mathbf{U}} + \mathbf{K}\mathbf{U} = \mathbf{F} \tag{1}$$

where **M** is the mass matrix, ***C*** is the damping matrix, ***K*** is the stiffness matrix, ***F*** is the load vector, ***U*** is the vector of displacements, $\dot{\mathbf{U}}$ is the vector of velocity, $\ddot{\mathbf{U}}$ is the vector of accelerations for each degree of freedom. For the undamped forced vibrations, the Equation (1) reduces to:

$$\mathbf{M}\ddot{\mathbf{U}} + \mathbf{K}\mathbf{U} = \mathbf{F} \tag{2}$$

The equation above was solved using the commercially available ANSYS implicit software and the Newmark's time integration scheme. In the case of a solid contact it was assumed that the contact layer between the adjacent wires has the properties of steel. In the case of a slip contact the layer between wires was simulated as solid, but with the shear wave velocity close to zero. In this case the shear wave is not transmitted between adjacent wires. The properties of two types of the contact layers between neighboring individual wires are presented in Table 1.

Just for simplicity not the whole structure of the strand of the multi-wire rope but only a row of five wires located on each other were analyzed (Figure 6). The diameter of a single wire in the model was 2 mm. The length of the simulated structure was 250 mm. Wires were meshed using SOLID45 elements which are used for modelling of solid structures. Each element had eight nodes with three degrees of freedom each: translation in x, y, and z directions. The average spatial size of the elements was 0.2 mm.

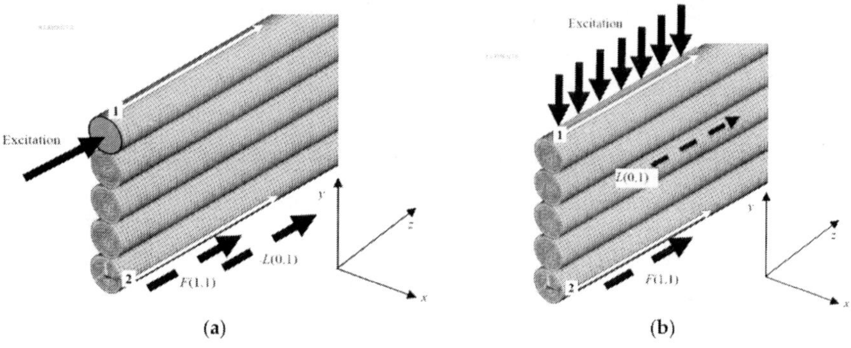

Figure 6. The sketch of FE model in the case of the edge type (symmetric) (**a**) and top (asymmetric) (**b**) excitation, when the diameter of a single wire is 2 mm and different types of the acoustic contact between neighboring wires was implemented: solid and slip. The numbers 1 and 2 denote the particular line for monitoring penetration of the guided wave into the structure.

For excitation of the UGW, the appropriate force for a three-period burst and the frequency of 60 kHz with the Gaussian envelope were applied in order to investigate wave propagation effects in more detail. Motivation of selection of 60 kHz frequency was based on the fact that, at a higher frequency, the wavelength of guided waves is two times shorter for the axial-like $L(0,1)$ mode and one and a half times shorter for the flexural-like $F(1,1)$ mode comparing to the case of 30 kHz. The interaction effects of propagating the $F(1,1)$-like mode with different types of the acoustic contact between the neighboring wires are stronger at the frequency of 60 kHz due to a lower ratio of displacements of radial and axial components. For the $L(0,1)$-like mode there is no significant difference due to a weak radial component. In addition, at mentioned both frequencies only the fundamental modes (e.g., the $L(0,1)$-like and the$F(1,1)$-like) propagate along the single thin wire and the influence of higher modes does not exists.

The symmetric mode in the upper wire was generated by applying a normal force to the edge of it covering the whole cross-section having the area of 3.3 mm^2 (Figure 6a). The asymmetric mode was generated applying the normal force to the top surface of the same wire, the length of the excitation region was 10 mm and the area of 10.45 mm^2 (Figure 6b). The bandwidth of the excitation pulse at −6 dB level was equal to 32 kHz. The integration step in the time domain was 0.5 μs. This step is 1/33 of the period in the case of the excitation pulse possessing the central frequency of 60 kHz.

The solution was obtained using numerical integration and in this way the particle velocity distribution in the whole simulated structure at different time instants was obtained. The results are presented in the form of B-scan images (Figure 7, Figure 8, Figure 9 and Figure 10), defining distribution of different components of a particle velocity along the central line of the upper surface of the first wire and the lower surface of the last (fifth) wire (Figure 6a,b).

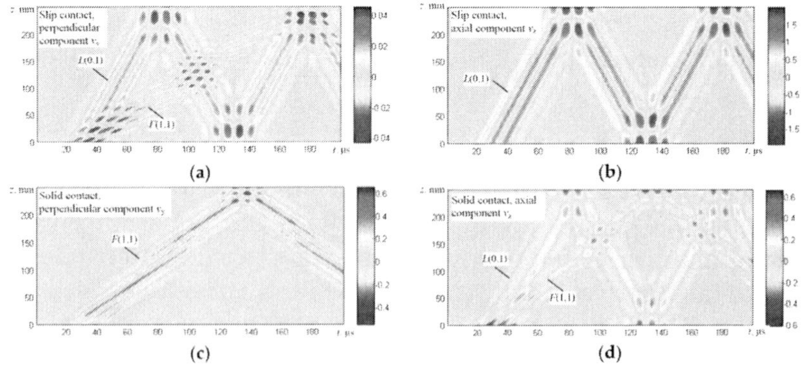

Figure 7. The B-scan images of the particle velocity along the first line of monitoring (the upper surface of the first wire) in the case of the edge type excitation: particle velocity components $v_y$perpendicular to the axis of the wire (**a**) and v_z (longitudinal-axial) (**b**) in the case of a slip contact and particle velocity components $v_y$perpendicular to the axis of the wire (**c**) and v_z (longitudinal-axial) (**d**) in the case of a solid contact between the neighboring wires.

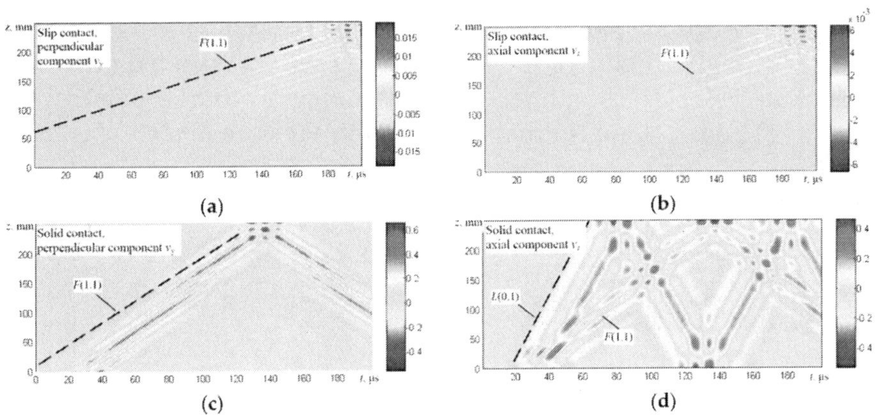

Figure 8. The B-scan images of the particle velocity along the second line of monitoring (the lower surface of the fifth wire) in the case of the edge type excitation: particle velocity components $v_y$perpendicular to the axis of the wire (**a**) and v_z (longitudinal-axial) (**b**) in the case of a slip contact and particle velocity components $v_y$perpendicular to the axis of the wire (**c**) and v_z (longitudinal-axial) (**d**) in the case of a solid contact between the neighboring wires.

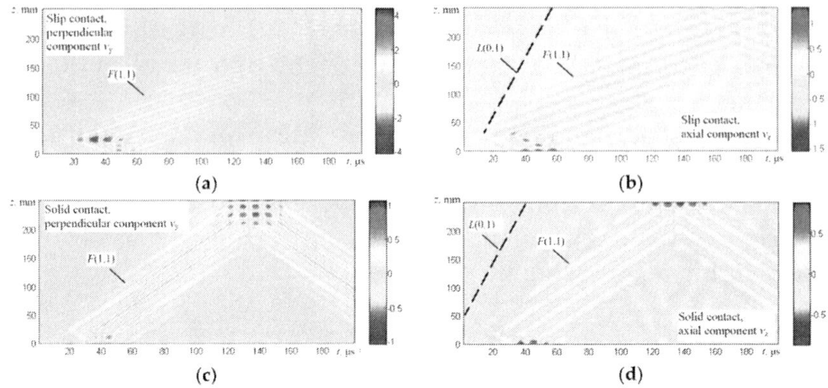

Figure 9. The B-scan images of the particle velocity along the first line of monitoring (the upper surface of the first wire) in the case of excitation from the top: particle velocity components $v_y$perpendicular to the axis of the wire (**a**) and v_z (longitudinal-axial) (**b**) in the case of slip contact and particle velocity components $v_y$perpendicular to the axis of the wire (**c**) and v_z (longitudinal-axial) (**d**) in the case of a solid contact between the neighboring wires.

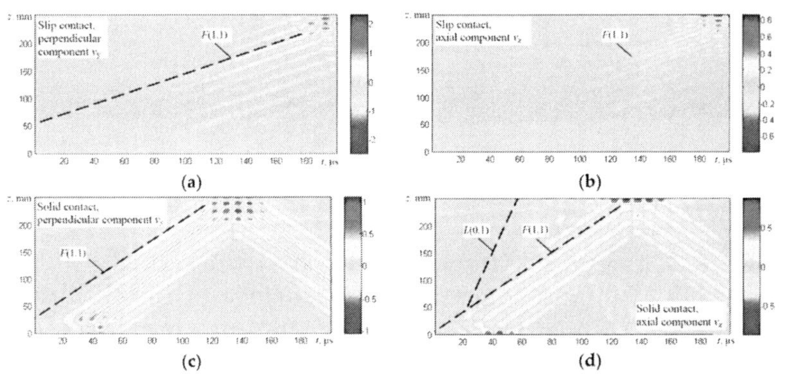

Figure 10. The B-scan images of the particle velocity along the second line of monitoring (the lower surface of the fifth wire) in the case of excitation from the top: particle velocity components $v_y$perpendicular to the axis of the wire (**a**) and v_z (longitudinal-axial) (**b**) in the case of the slip contact and the particle velocity components v_y perpendicular to the axis of the wire (**c**) and v_z(longitudinal-axial) (**d**) in the case of the solid contact between the neighboring wires.

It should be noted that the B-scan images have been acquired in a relatively long time interval, having a duration of 200 μs, and multiple reflections from edges of the simulated upper wire are clearly visible. The modelling results show that in the case of the edge type excitation and a slip contact between wires (Figures 7a,b and 8a,b) flexural $F(1,1)$ and symmetric $L(0,1)$ modes are generated in the first wire. However, only the flexural-like $F(1,1)$ mode after some delay penetrates to the bottom of the structure and can be observed on the lower wires as can be seen in Figure 8a,b. In the case of a solid contact between wires and edge type excitation (Figures 7c,d and 8c,d), not only the symmetric $L(0,1)$ mode, but the relatively strong asymmetric $F(1,1)$ mode is generated as well, and both of them penetrate the whole structure up to the last (fifth) wire (Figure 6a).

The results obtained in the case of the top side excitation by a normal force (Figure 6b) in general demonstrate that, in both cases, the dominant flexural-like $F(1,1)$ mode was generated which penetrates the whole structure (Figure 9 and Figure 10). In the case of a solid contact, the symmetric $L(0,1)$ mode is simultaneously generated also.

On the basis of the simulation results, it has been found that the phase velocities of the $F(1,1)$-like mode in the cases of solid and slip contacts are different. The analysis shows that in the case of the solid contact between wires, the phase velocity of the $F(1,1)$-like mode is around 1800 m/s. If the contact (during simulation) is assumed to be slip, the phase velocity decreases approximately by half, down to 900 m/s. This can be explained by the fact that, in the case of the slip contact, even the asymmetric guided wave mode $F(1,1)$-like propagates separately in each wire.

Therefore, in the case of the slip contact, there is no possibility to generate the $F(1,1)$-like mode along the overall structure of a multi-wire rope using the excitation (normal force) from the top effectively. In the case of the solid contact, the asymmetric mode $F(1,1)$-like corresponds to the fundamental mode$F(1,1)$-like of a much thicker simulated structure compared with the case of the slip contact.

Those results coincide well with the dispersion curves obtained using the analytical technique, which shows that the phase velocity of the $F(1,1)$ mode in a single wire should be around 900 m/s at the frequency of 60 kHz (Figure 2).

3. Finite Element Simulation of Guided Wave Propagation in 3D Wire Rope Model

The penetration of the propagating guided waves into the structure of the wire rope mainly depends on boundary conditions between the particular wires. During the finite element simulations it is usually assumed that the contact between the wires is solid to simplify the complexity of the numerical model. In this case, the surface-tangential component of the particle velocity can propagate throughout the structure of the rope. However, it is common that the multi-wire ropes are oiled to increase the flexibility and reduce the wear and abrasion. In this case, the penetration of guided waves may be quite different.

To investigate the effects of guided wave propagation and penetration into deeper layers of the multi-wire rope, the finite element simulations of the straight 3D structure were carried out. For investigations the multi-wire rope with the polypropylene core had been selected. The length of the wire rope was equal to 0.7 m. The simulations were performed for two separate cases. In Case A, the acoustic contact between separate straight strands was assumed to be solid and had the properties of steel. In Case B, the thin layer of a new material named as "elastic fluid" with the shear wave velocity close to zero was introduced between the separate straight strands to simulate a slip contact. The 3D views of the wire rope models for the Case A and the Case B are presented in Figure 11.

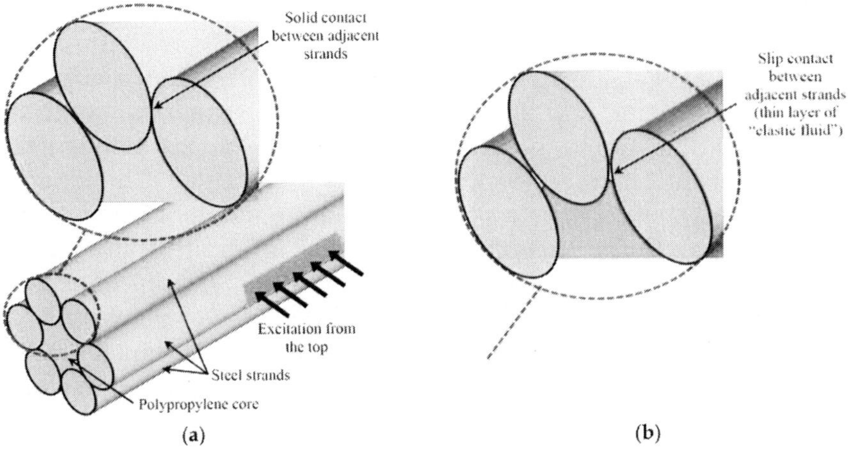

Figure 11. The 3D view of the multi-wire rope FE models in the case of a solid (**a**) and a slip (**b**) contact between neighboring strands.

The finite elements of the wire rope numerical model were created using a free mesh. The SOLID45 finite elements having eight degrees of freedom (translation in the *x*, *y*, and *z* directions) were used for meshing of the analyzed structure. The average spatial size of the elements was 2 mm for the steel region and 1.5 mm for the polypropylene. For the simulation of the layer of the "elastic fluid", the SOLID45 elements with the average spatial size of 1 mm were used. The number of nodes per wavelength of the slowest $F(1,1)$-like mode at 30 kHz frequency was 25 for steel, 33 for the polymer, and 50 for the elastic fluid.

In both cases the guided waves were generated by applying a normal force to the top surface of a separate strand. The excitation region had the dimensions of 15 mm × 10 mm and was placed 282 mm away from the nearest edge of the wire rope. This distance is equal to a single full whorl of the particular strand around the whole untwisted multi-wire rope.

In both cases the excitation pulse had the Gaussian envelope of seven periods and the frequency of 30 kHz. The bandwidth of the excitation pulse at the −6 dB level was 4.5 kHz. The Newmark time integration scheme was applied to obtain the numerical solution. The integration step in the time domain was 1.25 μs, which is 1/27 of the period at the 30 kHz central frequency. The spatial distributions of particle velocity components (axial v_z and perpendicular to the axis of the rope v_x) at different time instants were obtained using numerical integration. The C-scan images of the particle velocity distributions across the edge of the wire rope were created for the cases A and B. The C-scan images were created by selecting the peak-to-peak amplitude of the waveform at each available node of interest at time instants 55 μs, 81 μs, and 117 μs. Some further investigations demonstrated that the C-scan across the edge of the wire rope fully describes the spatial distribution of the wave amplitude.

The C-scan images of the axial v_z component (cross-section plane *y*-*x*) of the particle velocity obtained across the selected plane in the cases of a solid (case A) and a slip (case B) contacts are presented inFigure 12a–f. Please note that the amplitude scales in the C-scan images are different at different time instants and are increasing in time. The amplitudes at the time instant 117 μs are approximately 100 times larger than at 55 μs. The axial v_z component was selected for analysis in order to verify the obtained results by relevant experiments described further.

In the case of the solid acoustic contact mainly the flexural-like mode *F*(1,1), which propagates forward, is strongly expressed within the overall cross-section of the multi-wire rope (Figure 12a,c,e). In the case of the slip acoustic contact the *F*(1,1)-like mode propagates along the single strand only (Figure 12b,d,f).

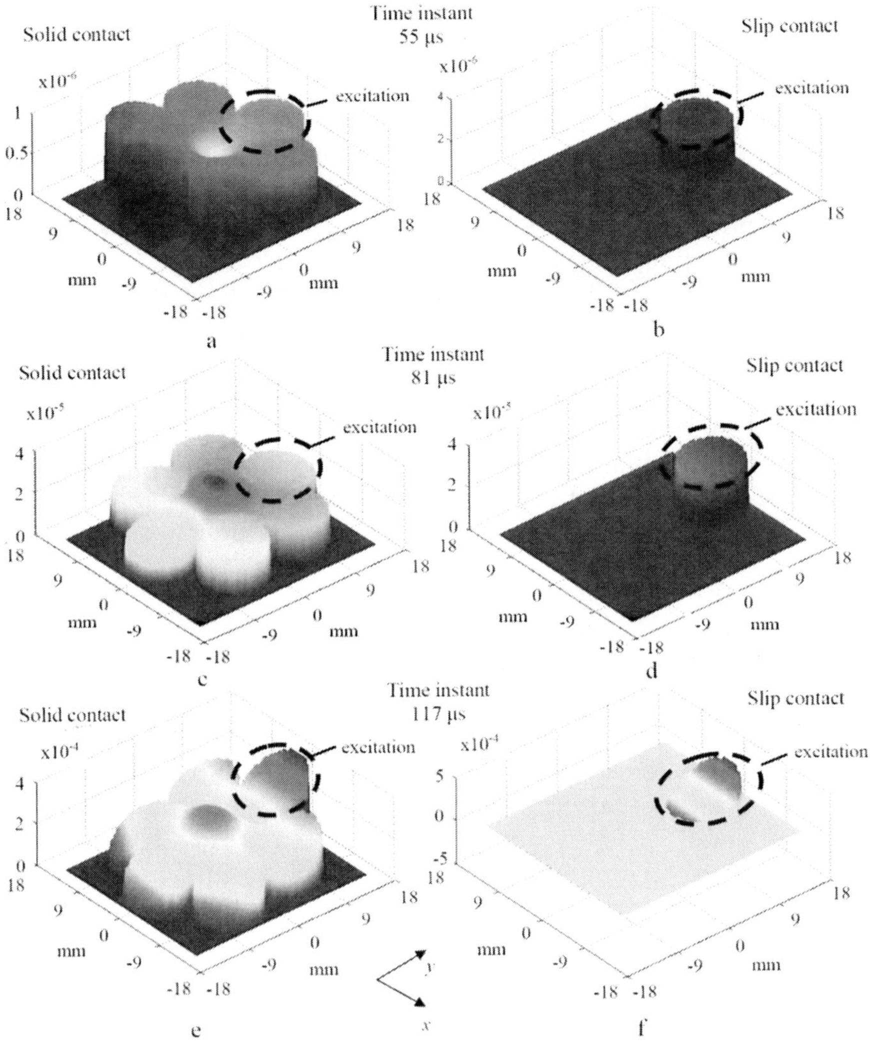

Figure 12. 3D representation of the axial component (v_z) of the particle velocity (cross-section plane *y-x*) at different time instants and acoustic contacts (solid or slip) between adjacent strands: time instant 55 μs ((**a**) solid contact, (**b**) slip contact), time instant 81 μs ((**c**) solid contact, (**d**) slip contact) and time instant 117 μs ((**e**) solid contact, (**f**) slip contact). The excitation was applied to the strand which is marked by a dashed line.

4. Experimental Investigation of a Multi-Wire Rope

From the numerical simulations it follows, that the penetration depth of the propagating UGW (mainly the flexural-like mode $F(1,1)$) into the internal structure of a multi-wire rope depends on the type of the acoustic contact between single wires and strands. The objective of the experimental investigation was to verify the effects of UGW propagation discovered by the FE simulation and to determine whether UGW propagate along the whole structure of wire rope or just in separate, externally-located strands.

The main purpose of the experiment was to reconstruct the wave penetration profile across the edge of the mock-up sample of the multi-wire rope. For the experimental investigation a composite rope with overall diameter of 35 mm and polymer core was selected (Figure 1).

In order to get the spatial distribution of the axial component amplitude across the edge of the multi-wire rope the air coupled technique was chosen. The immersion technique in this case is not suitable since the main goal of this experiment was to investigate the contact between separate strands. The water ingress can change the boundary conditions between the adjacent strands which would lead to inaccurate final results. The contact measurement technique was not suitable since it requires a stable acoustic force within the whole cross-section of the rope.

The experimental investigation of the wire rope was performed using the "Ultralab" ultrasonic measurement system, which had been developed at the Ultrasound Institute of Kaunas University of Technology. The experimental set-up for investigation of the multi-wire rope is presented in Figure 13. The ultrasonic guided waves were excited using the low frequency (f = 30 kHz, with a bandwidth of 20 kHz at the −6 dB level) ultrasonic transducers operating in the thickness mode (in order to create the normal force) and pitch-catch configuration. The transmitting transducer was attached to the surface of the particular strand using the specially developed spring-type adjusters. The contacting area of the conical delay line of the actuator of a concave shape was 134 mm² (projection of the contact area is the ellipse with the major radius of 8.6 mm and the minor radius of 5.8 mm). The transmitter was placed 272 mm away from the edge of the rope, which was equal to a single full whorl of the particular strand in the multi-wire rope. This distance is 10 mm shorter in comparison to the distance used in the finite element simulation (282 mm) as during the experiments the helical wire rope was used. To increase the transmission of the excitation force to the specimen a special conical delay lines with a concave contact surface were used. The transmitter was driven by the seven-cycle, 600 V tone burst with the central frequency of 30 kHz.

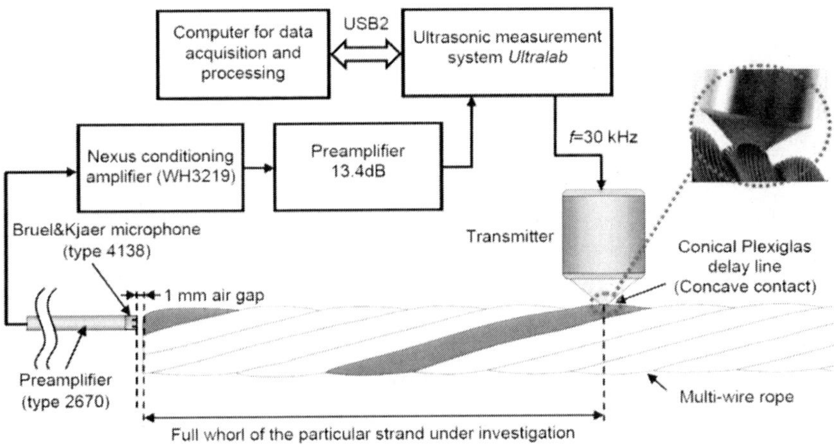

Figure 13. The experimental setup for the investigation of a guided wave propagation in a multi-wire rope.

The amplitude distribution across the edge of the rope was obtained by spatial scanning of a high frequency Brüel & Kjær air-coupled microphone (type 4138, Brüel & Kjær, Nærum, Denmark, frequency range: 6.5 Hz to 140 kHz, dynamic range: 52.2–168 dB, sensitivity 1 mV/Pa). The microphone was connected to the Falcon Range preamplifier (type 2670, Brüel & Kjær, Nærum, Denmark) and the Nexus conditioning amplifier (type WH3219, Brüel & Kjær, Nærum, Denmark). The diameter of the active region of the microphone was 3.175 mm, which is close to the diameter of a single wire of the rope. Before the experiments, the edge of the rope had been polished to get a precisely flat surface. The microphone was placed 1 mm from the edge of the rope, leaving as small air gap between the receiver and the specimen as possible. The microphone was operating in the near field, as the calculated near field length, taking into account the diameter of the particular strand and the wavelength of the longitudinal mode in the air, was equal to 2.22 mm. In general, the microphone makes it possible to register the sound pressure variations in air, across the edge of the multi-wire rope.

The sound pressure can be related to the particle velocity by the equation:

$$p(t) = \omega \times \xi \times Z = \frac{a(t) \times Z}{\omega} = \rho \times c \times v(t) \tag{3}$$

where ω is the angular frequency (rad/s), ξ is the particle displacement (m), Z is the acoustic impedance of air (N·s/m^3), $a(t)$ is the particle acceleration (m/s^2), ρ is the density of air (kg/m^3), c is the speed of sound (m/s) in air, and $v(t)$ is the particle velocity (m/s).

During the experiment, the peak-to-peak amplitudes of the axial component of the registered waveforms of the $F(1,1)$-like modes were measured within time interval of 0–250 μs. The selected relatively short time interval enabled us to distinguish between the signals propagating in the structure of the wire-rope and those which propagate directly to the microphone through the air gap. The experimentally-obtained distribution of the peak amplitudes of axial components across the tip of the multi-wire rope is presented in Figure 14. It is possible to conclude that the acoustic contact between adjacent strands of the real multi-wire rope is close to the slip type, according to the observed effect that UGW propagate mainly along the same strand on which the normal type excitation was applied. It corresponds well to the results obtained by the numerical FE simulation (Figure 12).

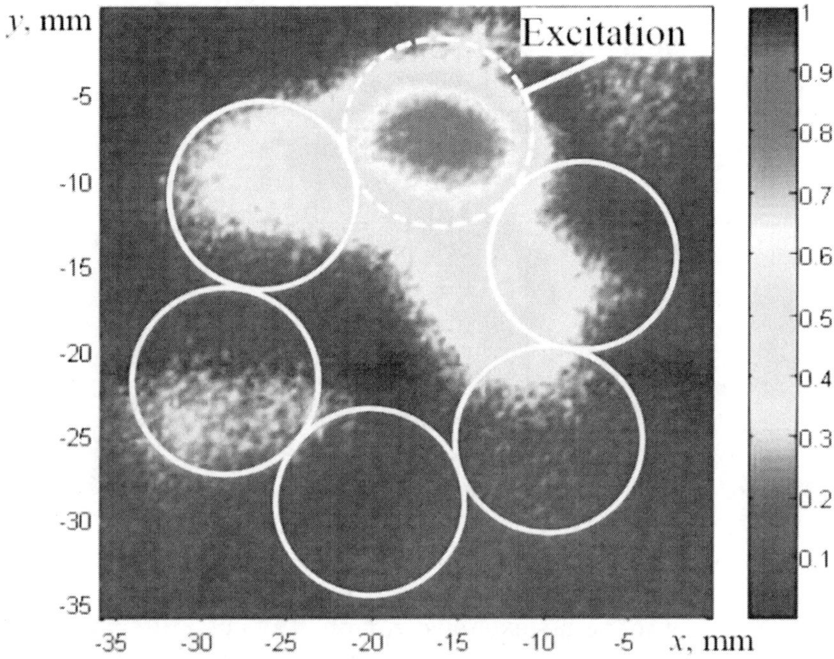

Figure 14. The experimentally-obtained C-scan image of the peak amplitudes of the axial components guided waves received across the cross-section of the multi-wire rope using the air coupled measurement technique. The excitation from the top was applied to the strand which is marked by a dashed line.

Previous investigations performed by contact type measurements of peak amplitudes of UGW (the*F*(1,1)-like mode) around the same strand and different strands proved that the highest amplitude of the received signal is registered when excitation and reception are performed on the same strand as well [23].

5. Conclusions

The important question regarding the development of the UGW-based testing technique has been the estimation of the real boundary conditions (solid or slip) between neighboring wires or strands and the real depth of penetration of UGW into the internal structure of the multi-wire rope. The propagating modes (mainly the $L(0,1)$-like and the $F(1,1)$-like) of UGW along the composite multi-wire rope have been identified by modelling using analytical, SAFE, and FE techniques. It has been estimated that some parts of the composite multi-wire rope (e.g., strands) may be replaced by a solid rod of the corresponding diameter. Otherwise, dimensions of the model are limited by a lack of computational resources and memory overflow.

During simulation of the row consisting of five wires placed on each other the obtained results have demonstrated that in the case of the edge type excitation (parallel to the axis of the wire) and the solid acoustic contact between adjacent wires, mainly the $L(0,1)$-like mode is generated. However, in the case of the solid contact, the $F(1,1)$-like mode is also generated due to mode conversions at each junction line between neighboring wires. In the case of excitation from the top by a normal force (perpendicular to the axis of the wire being excited), mainly the $F(1,1)$-like mode is generated for both types of the acoustic contact (slip and solid). The weak trace of the $L(0,1)$-like mode is also visible due to mode conversions.

The influence of different excitation regions and types of the acoustic contact between neighboring strands on the propagating UGW modes has been investigated using 3D FE modelling. Simulation results have demonstrated that in the case of the slip contact between adjacent strands and using the excitation from the top by a normal force the asymmetric-like mode $F(1,1)$ propagates separately in each strand. Therefore, in the case of the slip contact the possibility to generate efficiently the $F(1,1)$-like mode along the overall structure of a multi-wire rope is limited.

From the performed numerical modelling and relevant experiments it follows that the acoustic contact between adjacent strands of the real multi-wire rope is close to slip, according to the observed effect that UGW propagate mainly along the same strand on which excitation by a normal force has been applied. Investigations have demonstrated also that even in the case of the slip contact between wires the UGW can be used for integrity analysis (e.g., non-destructive testing) of particular strands of multi-wire ropes and to detect the defective ones.

The industrial demand for this work covers application of the developed ultrasonic pitch-catch testing technique, experimental set-up and special contact-type transducers to identify the particular defective strands in a multi-wire rope.

Acknowledgments

The authors would like to thank PhD student A. Jankauskas for provided help with experiments.

Author Contributions

Renaldas Raisutis, Rymantas Kazys and Liudas Mazeika conceived and designed the experiments, contributed in writing; Vykintas Samaitis and Egidijus Zukauskas performed numerical simulations, experiments and data analysis.

Conflicts of Interest

The authors declare no conflict of interest.

References

1. Yamasaki, T.; Tamai, S.; Hirao, M. Optimum excitation signal for long-range inspection of steel wires by longitudinal waves. *NDT E Int.* **2001**, *34*, 207–212.

2. Liu, Z.; Zhao, J.; Wu, B.; Zhang, Y.; He, C. Configuration optimization of magnetostrictive transducers for longitudinal guided waves inspection in seven-wire steel strands. *NDT E Int.***2010**, *43*, 484–492.

3. Cheong, Y.M.; Kim, S.; Jung, H.K. Effect of an alternating excitation and the applicability of a magnetostrictive transducer for a long-range guided waves inspection. In Proceedings of the 12th A-PCNDT 2006—Asia-Pacific Conference on NDT, Auckland, New Zealand, 5–10 November 2006; pp. 1–6.

4. Laguerre, L.; Treyssede, F. Non-destructive evaluation of seven-wire strand using ultrasonic guided waves. *Eur. J. Environ. Civ. Eng.* **2011**, *15*, 487–500.

5. Lanza di Scalea, F.; Rizzo, P.; Seible, F. Stress measurement and defect detection in steel strands by guided stress waves. *J. Mater. Civ. Eng.* **2003**, *15*, 219–227.

6. Kwun, H.; Bartels, K.A.; Hanley, J.J. Effects of tensile loading on the properties of elastic-wave propagation in a strand. *J. Acoust. Soc. Am.* **1998**, *103*, 3370–3375.

7. Laguerre, L.; Aime, J.C.; Brissaud, M. Magnetostrictive pulse-echo device for non-destructive evaluation of cylindrical steel materials using longitudinal guided waves. *Ultrasonics* **2002**, *39*, 503–514.

8. Rizzo, P.; Di Scalea, F.L. Wave propagation in multi-wire strands by wavelet-based laser ultrasound. *Exp. Mech.* **2004**, *44*, 407–415.

9. Treyssede, F. Elastic waves in helical waveguides. *Wave Motion* **2008**, *45*, 457–470.

10. Treyssede, F.; Laguerre, L. Wave propagation in helical multi-wire cables. In Proceedings of the 7th European Workshop on Structural Health Monitoring, Nantes, France, 8–11 July 2014; pp. 2044–2051.

11. Raišutis, R.; Kažys, R.; Žukauskas, E.; Mažeika, L.; Vladišauskas, A. Application of ultrasonic guided waves for non-destructive testing of defective CFRP rods with multiple delaminations.*NDT E Int.* **2010**, *43*, 416–424.

12. Gaul, L.; Sprenger, H.; Schaal, C.; Bischoff, S. Structural health monitoring of cylindrical structures using guided ultrasonic waves. *Acta Mech.* **2012**, *223*, 1669–1680.

13. Baltazar, A.; Hernandez-Salazar, C.D.; Manzanares-Martinez, B. Study of wave propagation in a multiwire cable to determine structural damage. *NDT E Int.* **2010**, *43*, 726–732.

14. Treyssede, F.; Laguerre, L. Investigation of elastic modes propagating in multi-wire helical waveguides. *JSV* **2010**, *329*, 1702–1716.

15. Hayashi, T.; Rose, J.L. Guided wave simulation and visualization by a semi-analytical finite element method. *Mater. Eval.* **2003**, *61*, 75–79.

16. Hayashi, T. Guided wave animation using semi-analytical finite element method. In Proceedings of the 16th World Conference on NDT, Montreal, QC, Canada, 30 August–3 September 2004; pp. 1–8.

17. Hayashi, T.; Song, W.L.; Rose, J.L. Guided wave dispersion curves for a bar with an arbitrary cross-section, rod and rail example. *Ultrasonics* **2003**, *41*, 175–183.

18. Bartoli, I.; Phillips, R.; Lanza di Scalea, F.; Salamone, S.; Coccia, S.; Sikorsky, C.S. Load monitoring in multiwire strands by interwire ultrasonic measurements. *Proc. SPIE* **2008**, *6932*, 1–12.

19. Marzani, A.; Viola, E.; Bartoli, I.; Lanza di Scalea, F.; Rizzo, P. A semi-analytical finite element formulation for modeling stress wave propagation in axisymmetric damped waveguides. *JSV***2008**, *318*, 488–505.

20. Laguerre, L.; Grimault, A.; Deschamps, M. Ultrasonic transient bounded-beam propagation in solid cylinder waveguide embedded in solid medium. *J. Acoust. Soc. Am.* **2007**, *121*, 1924–1934.

21. Beard, M.; Lowe, M.J.S.; Cawley, P. Ultrasonic guided waves for inspection of grouted tendons and bolts. *J. Mater. Civ. Eng.* **2003**, *15*, 212–218.

22. Kazys, R.; Butenas, G. Investigation of time-reversed ultrasonic waves focusing in multilayered media. *Ultragarsas* **2008**, *63*, 43–48.

23. Raisutis, R.; Kazys, R.; Mazeika, L.; Zukauskas, E.; Samaitis, V.; Jankauskas, A. Ultrasonic guided wave-based testing technique for inspection of multi-wire rope structures. *NDT E Int.* **2014**, *62*, 40–49.

© 2016 by the authors; licensee MDPI, Basel, Switzerland. This article is an open access article distributed under the terms and conditions of the Creative Commons Attribution (CC-BY) license (http://creativecommons.org/licenses/by/4.0/).

3
CHAPTER

Towards a Mathematical Model for Elastic Wave Propagation in Granular Materials

Leonardo Trujillo[1,2*], Vanessa Torres[1], Franklin Peniche[1,3], Leonardo Di G. Sigalotti[1]

[1]Centro de Física, Instituto Venezolano de Investigaciones Científicas (IVIC), Caracas, Venezuela

[2]The Abdus Salam, International Centre for Theoretical Physics (ICTP), Trieste, Italy

[3]Departamento de Física y Electrónica, Universidad de Cordoba, Montería, Colombia

ABSTRACT

A theoretical model for the propagation of acoustic waves in dry granular media is presented within the framework of the nonlinear granular elasticity. An essential ingredient is the dependence of the elastic moduli on compression. For the purpose of illustration, we analyze the case of a time-harmonic plane wave propagation under isotropic compression. We derive explicit relations for the wave speed dependence with the confining pressure. The present approach provides an accurate description of acoustic wave propagation in granular packings and represents a powerful tool to interpret the results of current experiments.

Keywords: Granular Media; Elastic Waves; Nonlinear Elasticity; Hydrostatic Compression; Granular Elasticity

1. Introduction

Granular materials consist of a collection of discrete macroscopic solid particles interacting via repulsive contact forces. Perhaps the simplest example of such a system is a dense packing of spherical glass beads under hydrostatic confining pressure. Such model systems are a useful starting point in the description of composite materials. Their physical behaviour involves complex nonlinear phenomena, such as non equilibrium configurations, energy dissipation and nonlinear elastic response [1]. We know from classical continuum dynamics that when a deformable body is under the action of a force distributed over its surface, the force is transmitted to every point inside the body [2]. Conversely, when we deal with a static packing of granular particles, the way by which the forces are transmitted within the packing remains a complex, and still unresolved, problem [3]. Several features of granular matter may be responsible for the complexity of the response to external perturbations. One important aspect lies on the observation that force networks form the skeleton that carries most of the load in a static granular medium [4]. The mechanical response of granular packings to external perturbations plays a major role in numerous engineering endeavours (such as soil mechanics and geophysics), as well as in industry (oil exploration, structural stability, product formulation in pharmacology and domestics, granular composites, heterogeneous materials, etc.).

Most laboratory experiments have been carried out using two-dimensional packings, consisting of photoelastic (i.e., birefringent under strain) disks, which have then allowed for a finest visualization of the generation and dynamical evolution of force chains. However, real granular materials are optically opaque, the photoelastic technique becomes difficult to practice. On the other hand, new tools such as pulsed ultrasonic transmission through granular beds under confined stress have been recently deviced to understand the elastic response of three-dimensional granular packings [5-7]. In particular, by studying the low-amplitude coherent wave propagation and multiple ultrasound scattering, it is possible to infer many fundamental properties of granular materials such as elastic constants and dissipation mechanisms [8].

The understanding of wave motion in granular media took a major step forward with the experimental observation of the coexistence of a coherent ballistic pulse travelling through an effective contact medium and a multiply scattered signal [5]. Such a picture is confirmed by the experiments of sound propagation in two dimensional regular lattice of spheres under isotropic compression [9], where the transmitted coherent signal remains almost unchanged for different packing realizations followed by an incoherent tail which depends on the specific packing configuration, in agreement with the completely disordered three dimensional case [5,6]. Besides the sensitive configuration coda type high frequency (HF) incoherent signals, one observes a primary low frequency (LF) coherent pulse [6,9]. This coherent pulse well defined at the leading edge of the transmitted signal corresponds to an effective wave propagating ballistically at compressional velocity $c \sim 1000$ m/s, and frequency $f \sim 70$ kHz. The sensitivity of the coherent and incoherent waves to changes in packing configurations has been studied in [6]. For independent granular samples, an ensemble average of configurations can cancel scattered signals and leave only the coherent wave. From experiments [6] one can infers that the coherent pulse is a self-averaging signal and configuration insensitive, which thus probes sound propagation in an equivalent effective contact medium. Usually the elastic response of a compressed granular packing is modeled as an assembly of spherical particles, where the contact forces between them are described by the Hertz law [10]. This model predicts that the speed of sound scales as 1/6 with the confining pressure P. However, experiments [6] (3D disordered packing) and [9] (2D regular lattice) show that for moderate pressures the speed of sound scales with a non-Hertzian exponent, roughly 1/4, followed by a Hertzian one at large values of the confining pressure.

In this paper we deal with a continuum description of wave propagation in granular packings, using the nonlinear elastic model developed in Refs [11-13]. This paper is structured as follows: In Section 2 we present a concise and detailed pedagogical introduction to granular elasticity; In Section 3 we introduce the equation of motion to describe the dynamical response of a granular packing subjected to an external perturbation and it is verified that the theory is consistent with classical elasticity; In Section 4 we solve the equation of motion for time-harmonic plane wave propagation under hydrostatic compression. Then, we derive a mathematical expression for the longitudinal wave speed with pressure dependence. In Section 5 we summarize the relevant conclusions.

2. Granular Elasticity

A basis for the construction of a theory for elastic wave propagation in granular media is provided by the elastic theory conveniently modified to account in infinitesimal forces due to force-dependent internal contacts between grains. The goal of this section is to introduce the granular elastic theory proposed by Jiang and Liu [11,12], which emphasizes the role of intrinsic features of granular dynamics such as volume dilatancy, mechanical yield and anisotropies in the stress distribution. Another fundamental ingredient is that the theory takes into account a general form of the type of contacts between the grains. On the other hand, the formulation directly extends the Boussinesq modified theory of elasticity and removes their thermodynamic inconsistency via a correct formulation of the strain free energy functional. In addition, under certain limits the theory includes the wellknown linear elasticity of isotropic and homogeneous elastic solids.

We begin with the basic notions used in the construction of a self-contained elastic theory for granular materials. To illustrate how a granular packing can be described by an elastic approach, we make the following assumptions:

- As a very first rule, we impose that the theory we will build should be invariant with respect to any Galilean transformation, i.e., any translation or rotation or combination of both, added to the displacement field of the medium should neither change their energy, nor any physical quantity we can derive from it such as the stress field.

- We do not consider the case for plastic deformations in the granular packing. Therefore, we assume that the predominant part of the energy is elastic. Elasticity implies reversible deformation, which change the internal energy ε of the deformed body according to $d\varepsilon = Tds + dW$, where T is the (thermal) temperature, S is the entropy and W the work due to the force exerted on the packing. We assume a reference equilibrium state a temperature T, from which we can construct an elastic energy potential (free energy functional) F. For a static granular media the granular temperature $T_g = 0$, which "measures" kinetic energy fluctuations when the system is fluidized. For a more detailed discussion and clarifications see references [11-13].

- As a mechanical model for the granular particles we consider smooth hard spheres with diameter d. We do not include friction at the contact surface between the spheres, and rotational degrees of freedom are neglected. Therefore, we consider the potential energy as a path-independent function.

- Congruently with the previous point, distributions of body or surface couples are not include in the present analysis. In this sense the present formulation is not complete and must be extended to include the effects of the grain rotation into the corresponding constitutive relations.

2.1. Continuum Theory Framework

The analysis of the deformation of a granular packing is rendered amenable to mathematical analysis by introducing the concept of a continuum medium. In this idealization it is assumed that properties averaged over a mesoscale are continuous functions of position and time. However, the presence of force chains at the grain scale, implying preferred force paths, have served as empirical argument against an isotropic continuum description of granular matter [3,4]. Nonetheless, recent findings on the stress distribution response to local and global perturbations have shed some light on the validity of using a continuum theory [14,15]. The passage from a microscopic to a macroscopic (continuum) mechanical description of granular and heterogeneous materials, including the mesoscopic disorder, has been recently addressed by Goldenberg and Goldhirsch [16-19]. They showed that exact continuum forms of the balance equations (for mass, momentum and energy) can be established as relations between spatially-weighted sums, in which a physical quantity $F(r)$ is approximated by the summation interpolant $F(r) \approx \sum_i F_i(r) W(r - r'|h)$, where $W(r - r'|h)$ is a smooth (differentiable) function, commonly referred as the interpolating kernel, and h is the smoothing (coarse-graining) length, which determines the spatial resolution and the scales for the spatial averages. These findings justify the continuum analysis adopted in the present work. In passing, we note that the approach followed in references [16-19] seems to be congruent with the Smoothed Particle Hydrodynamics (SPH) computational method; a meshfree particle method based on Lagrangian formulation, which has been widely applied to different areas in engineering and science [20].

2.2. Deformation: Displacements and Strains

The displacement field inside the granular packing will be characterized by a vector field u(r). Moreover, we require that the theory starts with assumption of small deformations, i.e., for an infinitesimal deformation the displacements field and their gradients are small compared to unity $(|u| \text{ and } |\nabla u| \ll 1)$. This choice limit the range of applicability of the theory but is physically reasonable for granular systems under strong static compression and small amplitude wave motion. However, the present model could provides the theoretical basis for analysis of ultrasonic wave propagation in geotechnical engineering, soil mechanics and rock mechanics.

To comply with the requirement of translational invariance, we express the elastic energy of the granular packing as a function of the derivatives of the displacement field. Since the displacement is a vector field, the gradient is a second order tensor whose components i; j can be written $\partial u_j / \partial x_i$. We can split the gradient into a symmetric and antisymmetric tensor. The later will contain the rotational part of the field. Because of invariance under rotations, we require that the energy should not depend on the antisymmetric part of the gradient of displacement. On the other hand, the displacement associated with deforming the grains, stores energy reversible and maintains a static strain. Sliding and rolling lead to irreversible, plastic process that only heat up the system. So, the total strain tensor u_{ij} may be decomposed into elastic and plastic parts $u_{ij} = \epsilon_{ij} + \epsilon_{ij}^{p}$. In this work we limit our analysis to granular packings under strong static compression. Therefore, the granular energy is a function of the elastic energy alone and we can neglect the plastic strain contribution. Up to linear order, the strain tensor is

$$u_{ij} \approx \epsilon_{ij} = \frac{1}{2}\left(\frac{\partial u_j}{\partial x_i} + \frac{\partial u_i}{\partial x_j}\right).$$

Let us remark that in the present analysis we do not include non-affine deformations. This assumption should be valid for strong compressed granular packings far from the jamming transition [21].

2.3. Strain Energy

In order to have an intrinsic form for the energy, i.e. which does not depend on the basis chosen to express the displacement field, we need to introduce only invariants of the tensor $\hat{u}$. A second order tensor in three dimensions, has only three invariants:

$$u_I = u_{ij} = \mathrm{Tr}\left(\hat{u}\right); \quad u_{II} = u_{ij}\ u_{ji} = \mathrm{Tr}\left(\hat{u}^2\right);$$

and $u_{III} = u_{ij}\ u_{jl}\ u_{il} = \mathrm{Tr}\left(\hat{u}^3\right)$.

Here $\mathrm{Tr}(\cdot)$ denotes the trace. In order to build a linear theory, we have to introduce an expression which is quadratic in the strain field. Thus only the square of the first invariant, and the second invariant, together with the assumptions of Galilean invariance and thermodynamic equilibrium, can be used and we are naturally led to the general expression of the elastic potential energy, given by the Helmholtz free energy functional [2]:

$$F(\hat{u}) = \frac{1}{2}\left[\lambda \mathrm{Tr}(\hat{u})^2 + 2\mu \mathrm{Tr}(\hat{u}^2)\right]$$

where λ and μ are material-dependent constants which characterizes the rigidity of the solids and are called the Lamé coefficients. If $\hat{u}$ contains only off-diagonal components, like for instance a pure shear, the elastic constant that comes into play is $\mu = \mathrm{G}$, the shear modulus. In the case of an isotropic compression, $\hat{u}$ is proportional to the identity tensor. The elastic constant relating the pressure to the decrease of volume is now $K = \lambda + 2/3\,\mu$, the compressibility modulus. In a model for granular materials, the elastic moduli K and G must be assumed to be proportional to the volume compression $\Delta = 1 - \rho_0/\rho = -u_{ll}$, where ρ_0 is the density in the absence of external forces [11,12]. So we have

$$K = \tilde{K}\Delta^b, \quad G = \tilde{G}\Delta^a, \tag{1}$$

with $\tilde{K}, \tilde{G} > 0$ for Δ ≥ 0 and $\tilde{K}, \tilde{G} = 0$ for Δ = 0, so that the elastic moduli remain finite. The exponents a and b are related to the type of contact between the grains. That is, when $a = b = 0$ linear elasticity is recovered, whereas $a = b = 1/2$ implies Hertz contacts [10]. This formulation provides a much better approximation to granular elastic behavior in which we can specify the type of contact by suitably choosing the exponents a and b. The Helmholtz free energy in the Jiang-Liu model is [11,12]:

$$F_{JL} = \frac{1}{2}\left[\tilde{K}\Delta^b\right]u_{ll}^2 + \left[\tilde{G}\Delta^a\right]u_s^2, \tag{2}$$

where $u_s^2 = u_{ij}^0 u_{ij}^0$, with $u_{ij}^0 = u_{ij} - 1/2 u_{ll} \delta_{ij}$.

The free energy strain potential (2) is stable only in the range of strain values that keeps it convex. Therefore, Equation (2) naturally accounts for unstable configurations of the system, as the yield, which appears as a phase transition on a potential-strain diagram.

2.4. Stress Field and Equilibrium

We now introduce the conjugated field to the strain: the granular stress field. By definition, this stress field, which is also a second-order symmetric tensor field, σ_{ij} is related to the strain field by

$$\sigma_{ij} = \frac{\partial F_{JL}}{\partial u_{ij}},$$

$$= -K u_{ll} \delta_{ij} + 2G u_{ij}^0 - \frac{1}{\Delta}\left(\frac{1}{2} bK u_{ll}^2 + aG u_s^2\right) \tag{3}$$

where δ_{ij} is the Kronecker's delta. Galilean invariance implies that the stress field should not do any work during a rigid motion of the solid. This property can be used to derive the balance equation satisfied by the stress field, always valid even if the rheological behaviour considered is not linear. When the granular packing is subjected to an external force density f, the balance equations read

$$\nabla \cdot \hat{\sigma} + f = 0. \tag{4}$$

This equation have been checked in references [22,23], where the authors analyzed several features of static granular packings, such as the stress dip in sand piles, the stress distribution in silos and under point loads, and the variation with pressure of the elastic moduli.

3. Granular Elastodynamics

Now we are ready to formulate a mesoscopic elastodynamic theory for granular materials. Above we stated that the theory is founded on the assumption of a continuum (coarse grained) field description and an elastic theory for granular materials which takes into account the compression dependent elastic moduli.

Equation of Motion

We start with the equation of motion for the elastic displacement u at time t and postion r,

$$\rho \frac{\partial^2}{\partial t^2} u(r,t) = \nabla \cdot \hat{\sigma} + f(r,t). \tag{5}$$

From the Jiang-Liu elastic model the stress tensor (3) which, by the Hooke's law, is given in terms of the Lamé coefficients by

$$\hat{\sigma} = -\left\{ \left[1 + \frac{b}{2} \right] \lambda + \left[\frac{b}{3} + a \frac{u_s^2}{\Delta^2} \right] \mu \right\} (\nabla \cdot u) \hat{I} \\ -\mu \left[\nabla u + (\nabla u)^T \right] \tag{6}$$

where $\hat{I}$ denotes the unitary tensor and the superindex ^{T} means transposition. Inserting Equation (6) into Equation (5) and rearranging terms, the equation of motion is

$$\rho \frac{\partial^2}{\partial t^2} u(r,t) = \left\{ \left[1 + \frac{b}{2} \right] \lambda \right. \\ \left. + \left[2 + \frac{b}{3} + a \frac{u_s^2}{\Delta^2} \right] \mu \right\} \nabla \left[\nabla \cdot u(r,t) \right] \\ -\mu \nabla \times \nabla \times u(r,t) + f(r,t), \tag{7}$$

this expression indicates that the elastic wave have both dilatational $\nabla \cdot u(r,t)$ and rotational deformations $\nabla \times u(r,t)$. Let us remark that setting $a = b = 0$, as it would be for linear elasticity, Equation (7) reduces to

$$\rho \frac{\partial^2}{\partial t^2} u(r,t) = (\lambda + 2\mu)\nabla\left[\nabla \cdot u(r,t)\right] \\ -\mu\nabla \times \nabla \times u(r,t) + f(r,t). \tag{8}$$

The above equation further simplifies to the wellknown wave equation

$$\frac{\partial^2}{\partial t^2} u(r,t) - c_p^2 \nabla\left[\nabla \cdot u(r,t)\right] \\ -c_s^2 \nabla \times \nabla \times u(r,t) = \rho^{-1} f(r,t) \tag{9}$$

The terms $c_p \sqrt{(\lambda + 2\mu)/\rho}$ and $c_s = \sqrt{\mu/\rho}$ are the compressional wave speed, and shear or transverse wave speed, respectively. This proves that the present elastodynamic model includes the well-known linear elasticity of isotropic and homogeneous elastic solids.

4. Wave Propagation

4.1. Time-Harmonic Plane Wave Propagation under Hydrostatic Compression

Now we proceed to analyze the case of time-harmonic plane wave under compression. We shall assume a perturbation of the packing by a time-harmonic plane deformation given by

$$u^P = U \exp\left[i\left(k \cdot r - \omega t\right)\right], \tag{10}$$

where the vector U is the amplitude of the displacement along the plane of the wave (polarization), while k is the wave vector and ω is the pulsation. We will now consider the case for isotropic compression. In the unperturbed reference configuration the strain tensor is

$u_{ij} = u_{ll}\delta_{ij}$, i.e.,

$$\hat{u} = \begin{pmatrix} u_{11} & 0 & 0 \\ 0 & u_{22} & 0 \\ 0 & 0 & u_{33} \end{pmatrix}, \tag{11}$$

and the trace is $\mathrm{Tr}(u_{ij}) = (u_{11} + u_{22} + u_{33}) = -\Delta$. The traceless part, $u_{ij}^0 = u_{ij} + 1/3\Delta\delta_{ij}$, is

$$\hat{u}^0 = \begin{pmatrix} 2u_{11} - u_{22} - u_{33} & 0 & 0 \\ 0 & -u_{11} + 2u_{22} - u_{33} & 0 \\ 0 & 0 & -u_{11} - u_{22} + 2u_{33} \end{pmatrix}. \tag{12}$$

Therefore, the invariant u_s^2 is

$$u_s^2 = \frac{2}{3}\Delta - 2\left(u_{11}u_{22} + u_{11}u_{33} + u_{22}u_{33}\right). \tag{13}$$

Finally, from (3) the stress tensor is

$$\sigma_{ij} = \left[\left(1 + \frac{b}{2}\right)K + a\frac{u_s^2}{\Delta^2}G\right]\Delta\delta_{ij} - 2u_{ij}^0 G. \tag{14}$$

Now we consider the displacements parallel to the wave normal plane along the x_1 axis and amplitude $\tilde{u}$

$$\tilde{u}_{\parallel}^{P} = \begin{pmatrix} \tilde{u} \\ 0 \\ 0 \end{pmatrix} \exp\left[\mathrm{i}\left(k_1 x_1 - \omega t\right)\right]. \tag{15}$$

As a next step, we need to calculate the disturbances. Thus the strain tensor associated with the wave is

$$\hat{\tilde{u}} = \mathrm{i}k\xi \begin{pmatrix} 1 & 0 & 0 \\ 0 & 0 & 0 \\ 0 & 0 & 0 \end{pmatrix}, \tag{16}$$

where $\xi = \tilde{u}\exp\left[\mathrm{i}\left(k_1 x_1 - \omega t\right)\right]$. The volume compression is $\tilde{\Delta} = -\mathrm{i}k\xi$. On the other hand, the traceless part $\tilde{u}_{ij}^{0} = \tilde{u}_{ij} - \frac{1}{3}\tilde{u}_{ll}\delta_{ij}$ is

$$\hat{\tilde{u}}^{0} = \frac{1}{3}\mathrm{i}k\xi \begin{pmatrix} 2 & 0 & 0 \\ 0 & -1 & 0 \\ 0 & 0 & -1 \end{pmatrix}. \tag{17}$$

Now the invariant $\tilde{u}_s^2$ is

$$\tilde{u}_s^2 = \frac{2}{3}\mathrm{i}k\xi\left(4u_{11} + \Delta\right). \tag{18}$$

The components of the perturbed stress tensor are given by

$$\tilde{\sigma}_{ij} = \left[(1+b)\left(1+\frac{b}{2}\right)K - a(1-a)\frac{u_s^2}{\Delta^2}G \right]\tilde{\Delta}\delta_{ij} + a\frac{\tilde{u}_s^2}{\Delta}G\delta_{ij} - 2\left[\tilde{u}_{ij}^0 + a\frac{u_{ij}^0}{\Delta}\tilde{\Delta} \right]G, \tag{19}$$

which in matrix notation have the diagonal form $\hat{\tilde{\sigma}}_{ij} = \mathrm{diag}(\sigma_{11}, \sigma_{22}, \sigma_{33})$, where

$$\sigma_{11} = \left[(1+b)\left(1+\frac{b}{2}\right)K - \left(a(1-a)\frac{u_s^2}{\Delta^2} - a\frac{\tilde{u}_s^2}{\Delta\tilde{\Delta}} + \frac{2}{3}\frac{a}{\Delta}(2u_{11} - u_{22} - u_{33}) - \frac{4}{3} \right) G \right]\tilde{\Delta}, \tag{20}$$

$$\sigma_{22} = \left[(1+b)\left(1+\frac{b}{2}\right)K - \left(a(1-a)\frac{u_s^2}{\Delta^2} - a\frac{\tilde{u}_s^2}{\Delta\tilde{\Delta}} + \frac{2}{3}\frac{a}{\Delta}(2u_{22} - u_{11} - u_{33}) + \frac{2}{3} \right) G \right]\tilde{\Delta}, \tag{21}$$

$$\sigma_{33} = \left[(1+b)\left(1+\frac{b}{2}\right)K - \left(a(1-a)\frac{u_s^2}{\Delta^2} - a\frac{\tilde{u}_s^2}{\Delta\tilde{\Delta}} + \frac{2}{3}\frac{a}{\Delta}(2u_{33} - u_{11} - u_{22}) + \frac{2}{3} \right) G \right]\tilde{\Delta}. \tag{22}$$

We can verify that $\partial\sigma_{22}/\partial x_2 = \partial\sigma_{33}/\partial x_3 = 0$. Finally, using Equation (15) and (20) we can ensemble the equation of motion to obtain

$$\rho\omega^2\xi = \left[(1+b)\left(1+\frac{b}{2}\right)K - \left(a(1-a)\frac{u_s^2}{\Delta^2} - a\frac{\tilde{u}_s^2}{\Delta\tilde{\Delta}}\right.\right.$$
$$\left.\left.+\frac{2}{3}\frac{a}{\Delta}(2u_{11} - u_{22} - u_{33}) - \frac{4}{3}\right)G\right]k_1^2\xi. \quad (23)$$

As we are interested in the case for isotropic compression, the components for the strain are $u_{11} = u_{22} = u_{33} = 1/3u_0$, thus $u_s^2 = \tilde{u}_s^2 = 0$. Therefore, the resulting equation of motion is

$$\rho\omega^2\xi = \left[(1+b)\left(1+\frac{b}{2}\right)K + \frac{4}{3}G\right]k_1^2\xi. \quad (24)$$

The pulsation ω and the wave vector k_1 define the longitudinal wave speed, i.e.,

$$\mathrm{v}_p := \frac{\omega}{k_1}. \quad (25)$$

Furthermore, noting that $K = \tilde{K}\Delta^b$ and $G = \tilde{G}\Delta^a$, the longitudinal wave speed is

$$\mathrm{v}_p = \sqrt{\frac{1}{\rho}}\left[(1+b)\left(1+\frac{b}{2}\right)\tilde{K}\Delta^b + \frac{4}{3}\tilde{G}\Delta^a\right]^{1/2}. \quad (26)$$

4.2. Speed of Sound and Pressure Dependence

In Refs. [5] and [9] the dependence of v_p vs the applied external pressure P was analyzed. The experiments show a scaling dependence for v_p with P. In order to capture the dependence for the calculated longitudinal wave speed (26) with the pressure, we use the granular free-energy functional (2). The pressure is given by

$$P = \frac{\partial F_{JL}}{\partial \Delta}, = \left(1+\frac{b}{2}\right)\tilde{K}\Delta^{b+1} \tag{27}$$

Therefore, the compression Δ is:

$$\Delta = \left[\frac{1}{\left(1+\frac{b}{2}\right)}\right]^{\frac{1}{b+1}} \left(\frac{P}{\tilde{K}}\right)^{\frac{1}{b+1}}. \tag{28}$$

If $a = b$, Equation (26) can be written as

$$\mathrm{v}_p = c_p \left(\frac{P}{\tilde{K}}\right)^{\frac{b}{2(b+1)}}, \tag{29}$$

where c_p is defined as

$$c_p = \left[(1+b)\left(1+\frac{b}{2}\right)\frac{\tilde{K}}{\rho} + \frac{4}{3}\frac{\tilde{G}}{\rho}\right]^{1/2} \left[\frac{1}{\left(1+\frac{b}{2}\right)}\right]^{\frac{b}{2(b+1)}}. \tag{30}$$

The present theoretical analysis, let us to conclude that the longitudinal wave speed scales with the pressure as

$$\mathrm{v}_p \sim P^{\frac{b}{2(b+1)}}, \tag{31}$$

where b is given by the mechanics of the contact model.

According to experiments [5] and [9] we have the following two cases:

For $b = 1/2$ (Hertzian contacts):

$$\begin{aligned} \mathrm{v}_p(P)\big|_{b=\frac{1}{2}} &= 0.963\left[\frac{15}{8}\frac{\tilde{K}}{\rho_0} + \frac{4}{3}\frac{\tilde{G}}{\rho_0}\right]^{1/2}\left(\frac{P}{\tilde{K}}\right)^{\frac{1}{6}} \\ &= c_p^H\left(\frac{P}{\tilde{K}}\right)^{\frac{1}{6}} \sim P^{\frac{1}{6}} \end{aligned} \tag{32}$$

For $\mathrm{b} = 1$ (non-Hertzian contacts):

$$\begin{aligned} \mathrm{v}_p(P)\big|_{b=1} &= 0.904\left[3\frac{\tilde{K}}{\rho_0} + \frac{4}{3}\frac{\tilde{G}}{\rho_0}\right]^{1/2}\left(\frac{P}{\tilde{K}}\right)^{\frac{1}{4}} \\ &= c_p^{nH}\left(\frac{P}{\tilde{K}}\right)^{\frac{1}{4}} \sim P^{\frac{1}{4}} \end{aligned} \tag{33}$$

Equations (32) and (33) behave according to the experiments in a qualitative manner. In **Figure 1**we compare the above equations with the data reported in [5].

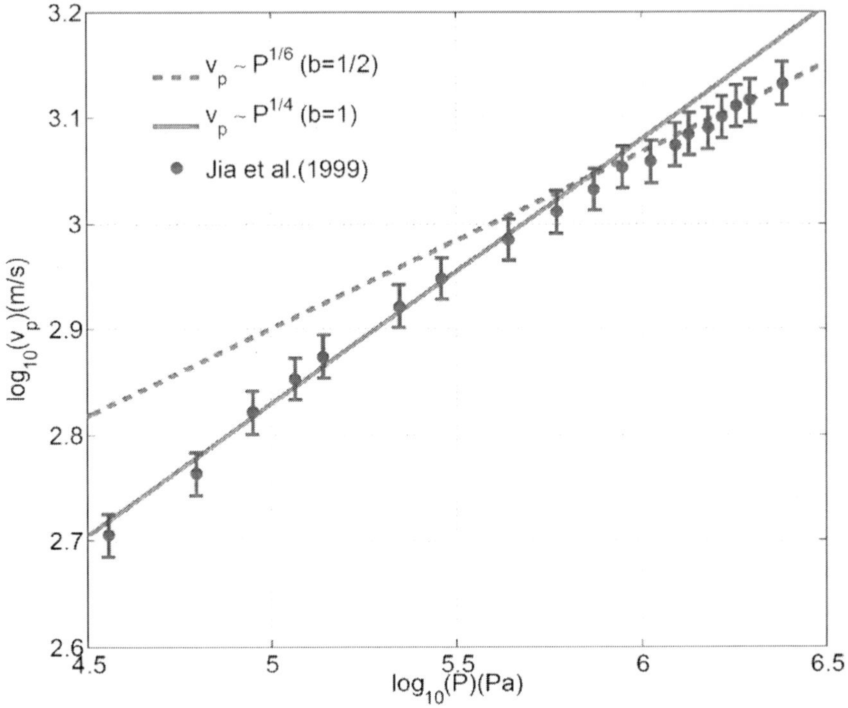

Figure 1. Velocity sound speed v_p (data points [5]) of the coherent wave versus the applied stress P. Red line: b = 1/2 (Hertzian contacts) with c_p^H = 7.9; Green dashed line: b = 1 with c_p^{nH} = 4.9.

However, at this point we have not a criterium to understand why the velocity "changes" with P between Equations (32) and (33), i.e., from the non-Hertzian contact model $b=1$ (33) for moderate pressures, to the Hertzian contact model $b=1/2$ (32) for higher pressures.

From the experimental data one can characterize the pressure dependence of the sound velocity introducing an effective exponent defined by the logarithmic derivative between v_p and P

$$\nu = \frac{d \ln v_p}{d \ln P}. \tag{34}$$

The compression Δ clearly influences the behavior of the velocity because it depends on the pressure. From Equation (29) we can estimate the dependence on Δ as

$$\left(\frac{v_p}{c_p}\right)^{\frac{2}{b}} = \Delta\left[\frac{1}{\left(1+\frac{b}{2}\right)}\right]^{-\frac{1}{b+1}} = \left(\frac{P}{\tilde{K}}\right)^{\frac{1}{b+1}}. \tag{35}$$

The qualitative behavior of ν can be obtained by interpolating from the asymptotic behavior at large and moderate pressures. This opens the question about if one can assume that ν is continuos, monotonically decreasing function from $\nu = 1/4$ to $\nu = 1/6$, while P increases. Or if it corresponds to discontinuous transition between the exponents.

Now we are in position to relate the parameter b and the exponent ν. We can estimate that b behaves according to the effective exponent ν as

$$b \sim \frac{\nu}{\frac{1}{2} - \nu}. \tag{36}$$

Verifying from the experiments, that the high pressure limit $\nu = 1/6$, implies for our model that $b \sim 1/2$. While for moderate pressures $\nu \sim 1/4 \Rightarrow b \sim 1$. The first case clearly corresponds to the Hertz contact model in the Jiang-Liu theory. The resistance to an external compression by a granular packing is due macroscopic rigidity mediated by the small contact regions between the grains and the contact areas increase with pressure. For spheroidal objects the Hertz model implies that the amount of deformation h increases with P as $h \sim P^{\frac{2}{3}}$ and the elastic energy U_H of the deformed sphere behave as $U_H \sim h^{\frac{5}{2}}$. This corresponds to the scaling at the high pressure regime. Various interpretations have been proposed for non-Hertzian contacts (e.g., [24,25]). However, let us remark that in 1971 Johnson, Kendal and Roberts [26] quoted some experimental outcomes where for moderated loads contacts areas between the bodies were considerably larger than those predicted by the Hertz model, while for high loads the results closely fitted the Hertz theory. Their observation is congruent with the scaling transition reported in [5] and [9]. The JKR theory [26] introduces the effects due to the energy at the surfaces of the contact bodies. According to [26], at moderate pressures tensile forces are present at the deformed surface modifying the Hertz contact law. They proposed that the tensile forces act as an attractive surface force even for clean and dry surfaces. These surfaces forces are of little significance at high loads, but become more significative as the pressure is reduced.

5. Conclusions

In this paper, we have shown that the nonlinear elastic theory proposed by Jiang and Liu [11] can be used to derive a time-evolution equation for the displacement field. The present formulation naturally accounts for features that are intrinsic to the dynamics of granular matter, considering microstructure information through the type of contact between the grains. The theory shows that the wave speed scales with pressure as $v_p \sim P^{\frac{b}{2(b+1)}}$. From the experimental data we have introduced the effective exponent υ and their relationship with the parameter b. This was done to characterize the transition from the non-Herztian regime $P^{1/4}$ to the Herztian regime $P^{1/6}$. However, we can not conclude if υ is a continuos, monotonically decreasing function of P, or a discontinuous transition at some critical value of P_c given by the intersection between Equations (32) and (33). On the other hand, we have provided an interpretation for the non-Hertzian contacts based on the JKR theory [26]. We hope that the present theoretical findings will stimulate further experimental works to explore in detail the behavior of contact deformations under compression and the effects on wave propagation in granular materials.

The mathematical model reported here is a first step towards the implementation and validation of computational engineering simulations for a variety of problems related to elastic wave propagation in granular composites. The equations of motion (7) are three partial differential equations which can be solved numerically for u_1, u_2 and u_3 provided with appropriate boundary and initial conditions, as for example using finite-element codes. In addition, the mathematical formulation at hand could help to inspire new ideas towards the construction of working theories for granular media, which may ultimately improve our interpretation of present experiments. For instance, understanding how waves do actually propagate in heterogeneous media is relevant to well-known problems such as earthquakes and seismic wave propagation and reflection in order to estimate the hydrocarbon content of potential oil wells.

REFERENCES

1. H. M. Jaeger, S. R. Nagel and R. P. Behringer, "Granular Solids, Liquids, and Gases," Review of Modern Physics, Vol. 68, No. 4, 1996, pp. 1259-1273.

2. L. D. Landau and E. M. Lifshitz, "Theory of Elasticity," Pergamon Press, New York, 1970.

3. S. Luding, "Granular Media: Information Propagation," Nature, Vol. 435, No. 7039, 2005, pp. 159-160.

4. T. S. Majmudar and R. P. Behringer, "Contact Force Measurements and Stress-Induced Anisotropy in Granular Materials," Nature, Vol. 435, No. 7045, 2005, pp. 1079- 1082.

5. X. Jia, C. Caroli and B. Velicky, "Ultrasound Propagation in Externally Stressed Granular Media," Physical Review Letters, Vol. 82, No. 9, 1999, pp. 1863-1866.

6. X. Jia, "Codalike Multiple Scattering of Elastic Waves in Dense Granular Media," Physical Review Letters, Vol. 93, No. 15, 2004, pp. 154303: 1-4.

7. X. Jia, J. Laurent, Y. Khidas and V. Langlois, "Sound Scattering in Dense Granular Media," Chinese Science Bulletin, Vol. 54, No. 23, 2009, pp. 4327-4336.

8. T. Brunet, X. Jia and P. Mills, "Mechanisms for Acoustic Absorption in Dry and Weakly Wet Granular Media," Physical Review Letters, Vol. 101, No. 13, 2008, pp. 138001: 1-4.

9. B. Gilles and C. Coste, "Low-Frequency Behavior of Beads Constrained on a Lattice," Physical Review Letters, Vol. 90, No. 17, 2003, pp. 174302: 1-4.

10. H. Hertz, "Über die Berührung Fester Elastischer Körper," Journal of Reine Angewandte Mathematik, Vol. 92, No. 1, 1882, p. 56171.

11. Y. Jiang and M. Liu, "Granular Elasticity without the Coulomb Condition," Physical Review Letters, Vol. 91, No. 14, 2003, pp. 144301: 1-4.

12. Y. Jiang and M. Liu, "A Brief Review of 'Granular Elasticity'," European Physical Journal E, Vol. 22, No. 3, 2007, pp. 255-260.

13. Y. Jiang and M. Liu, "Granular Solid Hydrodynamics," Granular Matter, Vol. 11, No. 3, 2009, pp. 139-156.

14. D. Serero, G. Rydellet, E. Clément and D. Levine, "Stress Response Function of a Granular Layer: Quantitative

Comparison between Experiments and Isotropic Elasticity," European Physical Journal E, Vol. 6, No. 2, 2001, pp. 169-179.

15. W. G. Ellenbroek, M. van Hecke and W. van Saarloos, "Jammed Frictionless Disks: Connecting Global and Local Response," Physical Review E, Vol. 80, No. 6, 2009, pp. 061307: 1-18.

16. C. Goldenberg and I. Goldhirsch, "Force Chains, Microelasticity and Macroelasticity," Physical Review Letters, Vol. 89, No. 8, 2002, pp. 0843021: 4.

17. I. Goldhirsch and C. Goldenberg, "On the Microscopic Foundations of Elasticity," European Physical Journal E, Vol. 9, No. 3, 2002, pp. 245-251.

18. C. Goldenberg and I. Goldhirsch, "Small and Large Scale Granular Statics," Granular Matter, Vol. 6, No. 2-3, 2004, pp. 87-96.

19. C. Goldenberg and I. Goldhirsch, "Friction Enhances Elasticity in Granular Solids," Nature, Vol. 435, No. 7039, 2005, pp. 188-191.

20. G. R. Liu and M. B. Liu, "Smoothed Particle Hydrodynamics: A Meshfree Particle Method," World Scientific, Singapore City, 2003.

21. T. S. Majmudar, M. Sperl, S. Luding and R. P. Behringer, "Jamming Transition in Granular Systems," Physical Review Letters, Vol. 98, No. 5, 2007, pp. 058001: 1-4.

22. D. O. Krimer, M. Pfitzner, K. Bräuer, Y. Jiang and M. Liu, "Granular Elasticity: General Considerations and the Stress Dip in Sand Piles," Physical Review E, Vol. 74, No. 6, 2006, pp. 061310: 1-10.

23. K. Bräuer, M. Pfitzner, O. Krimer, Y. Jiang and M. Liu, "Granular Elasticity: Stress Distributions in Silos and Under Point Loads," Physical Review E, Vol. 74, No. 6, 2006, pp. 061311: 1-10.

24. J. D. Goddard, "Nonlinear Elasticity and Pressure-Dependent Wave Speeds in Granular Media," Proceedings of the Royal Society of London A, Vol. 430, No. 1878, 1990, pp. 105-131.

25. P.-G. de Gennes, "Static Compression of a Granular Medium: The 'Soft Shell' Model," Europhysics Letters, Vol. 35, No. 2, 1996, pp. 145-149.

26. K. L. Johnson, K. Kendall and A. D. Roberts, "Surface Energy and the Contact of Elastic Solids," Proceedings of the Royal Society of London A, Vol. 324, No. 1558, 1971, pp. 301-313.

4

CHAPTER

Propagation and Confinement of Electric Field Waves along One-Dimensional Porous Silicon Hybrid Periodic/Quasiperiodic Structure

José Escorcia-García[1], Miguel Eduardo Mora-Ramos[2,3]

[1]Centro de Investigación en Energía, Universidad Nacional Autónoma de México, Temixco Morelos, México

[2]Física Teórica y Aplicada, Escuela de Ingeniería de Antioquia, Sede Las Palmas, Envigado Antioquia, Colombia

[3]Facultad de Ciencias, Universidad Autónoma del Estado de Morelos, Cuernavaca Morelos, México

ABSTRACT

Selective spatial confinement of the electric field intensity is theoretically obtained for the light propagation along hybrid structures conformed by periodic and Fibonacci quasiperiodic dielectric multilayers. Sandwich-like configurations featuring periodic-quasiperiodic-periodic as well as quasiperiodic-periodic-quasiperiodic designs exhibit spatial localization of a large percent of the optical signal within specific zones of the hybrid system. Such a feature might be of interest in the pursuing of lasing devices based on porous silicon. It is found that the electric field confinement does not only depend on the quality of the defect or a particular transmission mode observed in the reflectivity spectra. We show that it is possible to enhance the electric field confinement solely varying the angle of incidence. The possibility of realizing finite photonic crystals with reduced size and very well defined band gap by means of a quasiperiodic-periodicquasiperiodic hybrid multilayer is also revealed.

Keywords: Optical Properties; Electric Field; Multilayers; Hybrid Structures

1. Introduction

The engineering of electromagnetic modes at optical frequencies in artificial dielectric structures, with periodic, quasiperiodic or random variation in the refractive index, was first proposed independently by Yablonovitch, Kohmoto and John in 1987 [1-3]. Such kinds of complex dielectrics are often considered as photonic crystals or quasicrystals, and can be much easily obtained in a onedimensional multilayered configuration. Their attributes have generated worldwide research and development of sub-μm and μm size active and passive photonic devices such as a single-mode and non-classical sources, guided wave devices, resonant cavity detection, and components for optical communication.

Quasicrystals are non-periodic structures constructed following a simple deterministic generation. Among the different structural sequences used for quasicrystal fabrication one finds Cantor-like, Fibonacci (FN), generalized Fibonacci, Thue-Morse (TM), generalized ThueMorse, Period Doubling and Rudin-Shapiro. Although the FN system has been the subject of most study in this field. Electronic properties of FN multilayer structures are quite widely investigated (see for instance [4-7], and references therein). There are also studies on vibrational frequencies and atom displacements [8], as well as the propagation and localization of elastic waves [5,9-11]. Although less investigated than FN ones, the TM multilayered systems have also deserved significant attention. Their electronic properties have been studied, for instance, in [4,6,12,13]. In addition, some properties of elastic waves in quasiregular TM structures are reported in [14,15].

From the optical point of view, previously done work has focused on the study of the localization of light waves within FN quasiperiodic multilayer quasicrystals which leads to the appearance of photonic bandgaps, in a similar way to those existing in periodic structures. The photonic localization in the dielectric microstructure bears an analogy to the electronic localization in crystals [3]. Later on the concept of localization was recognized as applicable to any type of wave, such as acoustical wave [16]. Kohmoto et al. introduced the first system based on optical FN multilayers capable of localizing photons [2]. In 1994 Gellermann et al. showed experimental proof of the existence of bandgaps in the spectrum of FN dielectric multilayers [17]. There are also reports in the literature about the obtention of omnidirectional bandgaps using FN quasiperiodic structures [18,19]. A study on the features of the optical propagation in porous silicon-based FN polytype multilayers was put forward by Agarwal and Mora-Ramos [20]. Optical properties of non-Fibonacci 1D dielectric quasiregular systems have also been studied by some authors [20-41].

On the other hand, in a couple of works on quasiperiodic dielectric multilayers, E. Maciá outlined the possibility of designing hybrid dielectric heterostructures [42, 43]. Soon after, Wen et al. put forward the realization of an omnidirectional reflector made of on the basis of a Fibonacci-Bragg(periodic) hybrid system [44]. Then, for some years, the study of hybrid dielectric systems was practically out of literature. It is worth mentioning that during such period A. Montalbán and his collaborators lead a research on the behavior of oscillation modes in hybrid 1D periodic/quasiregular chains of point masses [45-47]. One of the main results of these works is the appearance of a selective spatial confinement of the vibration amplitudes, whether it is within the multilayered periodic part or whether it takes place within the quasiregular region of the structure.

Motivated by those reports on mechanical oscillations, in 2009 we presented some preliminary theoretical reports on the optical propagation in porous silicon-based multilayers designed as hybrid combinations of FN and Bragg mirror (BM) dielectric sections [48-50]. In these works we obtained an enhancement of in-gap microcavity mode localization as well as of the omnidirectionality of the reflectance properties of the structures proposed. Moreover, a significant selective spatial localization was detected for some propagating modes. That is, there are wavelengths for which the intensity of the electric field confines almost totally within the FN section or within a BM one. Some refinements of these results, but still preliminary ones were published in a more recent article [51]. It is worth mentioning that, later than the publication of our 2009 reports there have been some communications in relation with optical properties of hybrid periodic/ quasiregular and quasiregular/quasiregular hybrid dielectric heterostructures [52-54]. There is also a very recent experimental and theoretical report on tunable resonant transmission in a kind of porous silicon-based heterostructures [55]. The main interest in these works is focused on their use as reflectors and filters, and nothing is said there about the confinement or localization of the optical waves.

In this article we are aimed at developing a more detailed study of the selective spatial localization of the electric field of light waves propagating along hybrid FN-BM heterostructures. The discussion will consider both FN-BM-FN and BM-FN-BM multilayer designs together with the effects of the angle incidence and the variation of the dielectric contrast by changing the refractive index of the two basic layers involved. The work is organized with a section that follows on, presenting the model and the calculation scheme. Then there is a section containing the analysis and discussion of the obtained results. Finally, the main conclusions of the work shall be given.

2. Model and Methodology

A hybrid structure is defined as a combination of two independent structures with different properties, in this case a periodic and an aperiodic structure, to form a structure with superior features. As we previously mentioned, for the periodic part we choose the classical BM structure, while for the aperiodic we chose the FN one. These structures are produced with the use of the following substitution rules: $A_j \rightarrow A_{j-1}B_{j-1}$, $B_j \rightarrow A_{j-1}B_{j-1}$ and $A_j \rightarrow A_{j-1}B_{j-1}$, $B_j \rightarrow A_{j-1}$ for the periodic and quasiperiodic structure, respectively. Both constituents have $A_0 = A$ and $B_0 = B$ as a seed. So, for instance, the third generation period structure is given by BM(3) = ABABABAB while the fourth generation Fibonacci structure is given by FN(4) = ABAABABA. Once they are constructed, the hybrid assembly design implies putting them together in a sandwich configuration, for which we are considering two different ways: BM-FN-BM and FN-BM-FN. In our particular case, we use the pair of refractive indices $n_A = 1.8$ and $n_B = 1.2$ which are commonly used in the fabrication of porous silicon heterostructures for photonic applications [20,34,36,39]. The choice for the layer widths is made according the quarter-λ rule: $d_{A,B} = \lambda_0/4n_{A,B}$, with $\lambda_0 = 800$ nm. The**Figure 1** schematically shows the BM-FN-BM design, with the indication of the incident, reflected and transmitted waves. Viewed in this way, the proposed structure can be considered as a system of two or more defects (depending upon the particular order of the FN generation) included in the otherwise periodic configuration. A bulky defect would be obtained if the central (FN) part were made of layers with values of the refractive index distinct to those used in the BM part. But this possibility will not be discussed here, rather it could be the subject of another study.

The calculation of the electric field of s and p polarizations in the hybrid dielectric multilayer uses the well known transfer matrix method. Its details can be found elsewhere [56-58]. The approach makes provision for the variation of the angle of incidence as well as for the discontinuity in the dielectric properties when passing from a layer A to a layer B and viceversa. Within this framework, we are able to calculate the reflection and transmission coefficients, but our main interest is the obtention of the relative electric field intensity (defined as the quotient $|E(z,\omega)/E_i|^2$, with E_i being the amplitude of the incident signal) as a function of the position z in the structure, measured from the point of light incidence. The expression that allows us to evaluate the electric field amplitude in terms of the transfer matrix is

$$\boldsymbol{E}(z,\omega)=\left[1+r(\omega)\right]w_{11}(z)+\gamma_0\left[1-r(\omega)\right]w_{12}(z)$$

where $w_{\alpha\beta}$ are the elements of the matrix that connects the field at the extreme left-hand interface with the layer in which the coordinate z lies. The quantity $r(\omega)$ is the complex reflectance of the whole structure, and the factor γ_0 appears as a result of the imposition of suitable boundary conditions on the field components at the interface of light incidence (see **Figure 1**). Then, the field intensity is simply the absolute square of the electric field, $|E|^2$, which is normalized to the amplitude of the incident wave.

3. Results

The **Figure 2** contains the calculated optical properties of hybrid Bragg-Fibonacci heterostructures in different configurations. The color density graphics correspond to the relative electric field intensity associated to the optical mode with energy hω (vertical axis), at a given position z in the multilayer structure. On the side insets, one can see the corresponding normal-incidence reflectance, $R = |r(\omega)|^2$, in each case.

In the graphics 2a and 2b we are representing the results for hybrid BM3-FN4-BM3, and aBM3-aFN4-aBM3 dielectric heterolayers. The prefix "a" indicates that the structure considered has the refractive indices of its layers

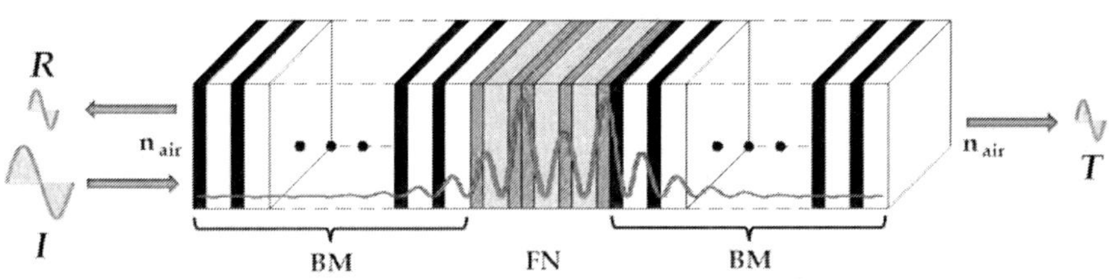

Figure 1. (color online) Schematic representation of the incidence and propagation of the electric field wave along a hybrid BM-FN-BM.

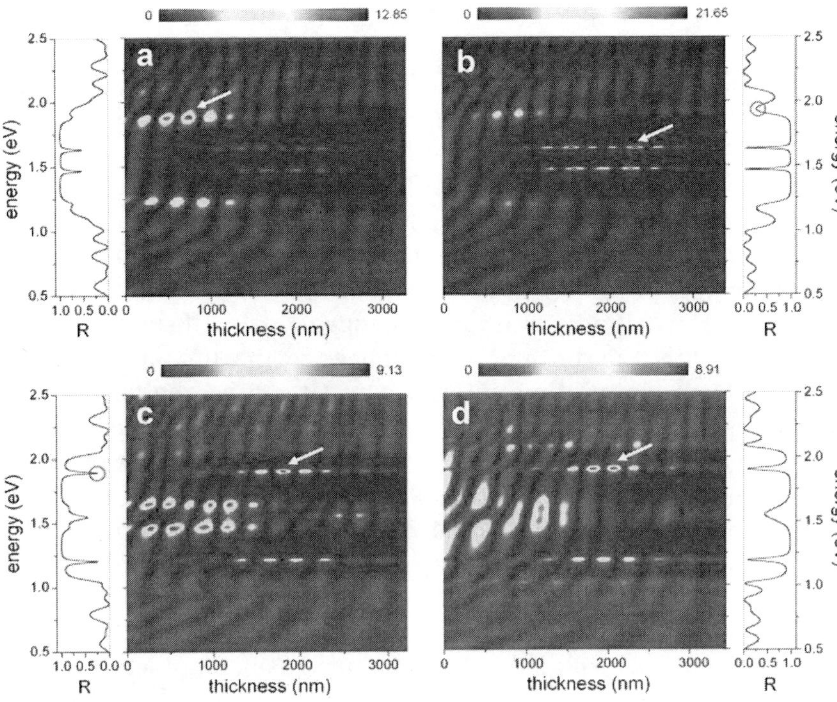

Figure 2. (color online) Distribution of the relative electric field intensity along the hybrid structures for (a) BM3-FN4-BM3, (b) aBM3-aFN4-aBM3, (c) FN4-BM3-FN4, and (d) aFN4-aBM3-aFN4 with the positions of the corresponding $|E|^2$ maxima whose values appear at the right of the color bars over the figures, indicated by arrows. The insets on the side of each graph gives the reflectivity at normal incidence of the particular structure.

exchanged; that is, $n_A \leftrightarrow n_B$. Both BM3 and FN4 parts consist of eight monolayers. On the other hand, the FN4 contains a total of five "A" layers whilst there are only four in the BM3. This makes the physical thickness of each part to have a different value. According to the $\lambda/4$ design, the BM3 and aBM3 parts have the same width: 1111 nm. However, the total thicknesses of the FN4 and aFN4 substructures are different: 1056 nm and 1167 nm, respectively. Color bars located above each graphics serve as intensity-scale indicators: from zero (deep blue) at the left to maximum (dark red) intensities at the right.

In each case, the maximum value of the relative intensity attained within the structures appears at the top right of the figure. The arrows are pointing at the positions of such maximum intensity peaks within the structures.

From the analysis of Figures 2(a) and 2(b) we can see that the relative field intensity of some particular modes show a selective spatial confinement, whereas the intensity of the electric field of the remaining propagating optical modes in the energy interval considered distributes rather homogeneously throughout the structure, with the normal occurrence of maxima and minima resulting from the multiple interference. Of course, the far lefthand part of the system shows the presence of illuminated zones even in the energy region within the photonic band gap. This is due precisely to its proximity to the surface of light incidence. The maximum relative intensity of the BM3-FN4-BM3 corresponds to a mode with energy immediately above the gap. The most of the spatial location of the associated electric field squared amplitude is approximately at the middle point of the left Bragg mirror. One readily sees that when $z > 1000$ nm, the relative intensity decreases significantly. It is possible to observe also the spatial localization of rather high field intensity in the left-hand Bragg multilayer. The energy of such a mode is now immediately below the gap. However the values of $|E|^2$ in this case are smaller. For these two optical modes, the confinement of a greater part of the electric field amplitude at the left of the system can be explained as follows: The remaining parts of the hybrid heterostructure are acting as mirrors that reflect the light back to the left (notice that the region containing the right-hand Bragg part is colored in darker blue which means that almost no light is propagating throughout it).

There are two signals of low, but nonzero relative intensity with spatial field localization mainly in the central FN4 substructure. These correspond to the pair of microcavity modes within the photonic band gap. Such modes can be related with the situation of a "double A-defect", introduced by the FN substructure. The same double microcavity is present in the aBM3-aFN4-aBM3 configuration**Figure 2**(b). They are at the same energy position of those of the BM3-FN4-BM3 structure; but from the reflection spectrum one may see that they are sharply defined, reaching almost the unity transmission. With regard to their spatial field dependence, we can say that the change in the dielectric contrast (with n_A now having the lower of the two values) has led to the change in the relative intensity localization. In this case the microcavity modes and not the bandgap-edge ones exhibit the strongest confinement, which takes place within the aFN4 region this time.

Figures 2(c) and 2(d) show the calculated relative intensity and total reflectance for hybrid FN4-BM3-FN4 and aFN4-aBM3-aFN4 multiple heterolayers. This is not the typical Bragg-microcavity optical complex design. Now the photonic band gap appears split into two parts separated by a region containing two main propagating modes, whose relative intensities confined mostly within the left-hand FN4 or aFN4 substructures. Interestingly, the process of multiple internal reflections in the hybrid system makes these two optical signals to travel along the structure in such a way that, close to its right-hand end, they seem to merge into a wave with a an intermediate frequency. This frequency is the one at which there is a local minimum of the reflectance in the separated transmission gap structure.

However, the highest relative field intensity in these structures corresponds to a microcavity-like mode located close to the high-frequency edge of the photonic band gap. In this case, the main spatial confinement of the field occurs in the central BM part of the hybrid multilayer, although non-vanishing field intensities can be observed in the two adjacent FN4 substructures as well. In all, the highest values of the relative intensities attained at the maxima indicated by the arrows, are smaller than those occurring in the BM3-FM4-BM3 structures of Figures 2(a) and 2(b).

The physical reason behind the behavior of the light propagation in the hybrid dielectric heterostructures is the process of multiple internal reflections associated to the interfaces that separate regions of different dielectric properties. The discussion made above assumes that the light incidence to be normal to the layers. However, the electromagnetic usually reaches the structure forming an angle with the normal to the layer's plane. Then, there will be multiple Snell-type refraction processes at the interfaces. Taking this into account, the **Figure 3** presents the spatial distribution of the calculated relative field intensity for particular optical modes as a function of the angle of incidence.

Figure 3(a) contains the relative field intensity of both the s and p polarizations of an optical wave signal with a frequency mode of 1.8955 eV propagating in the FN4- BM3-FN4 hybrid dielectric heterostructure considered above. In the case of the p polarization, the field intensity distributes rather homogeneously over all the spatial

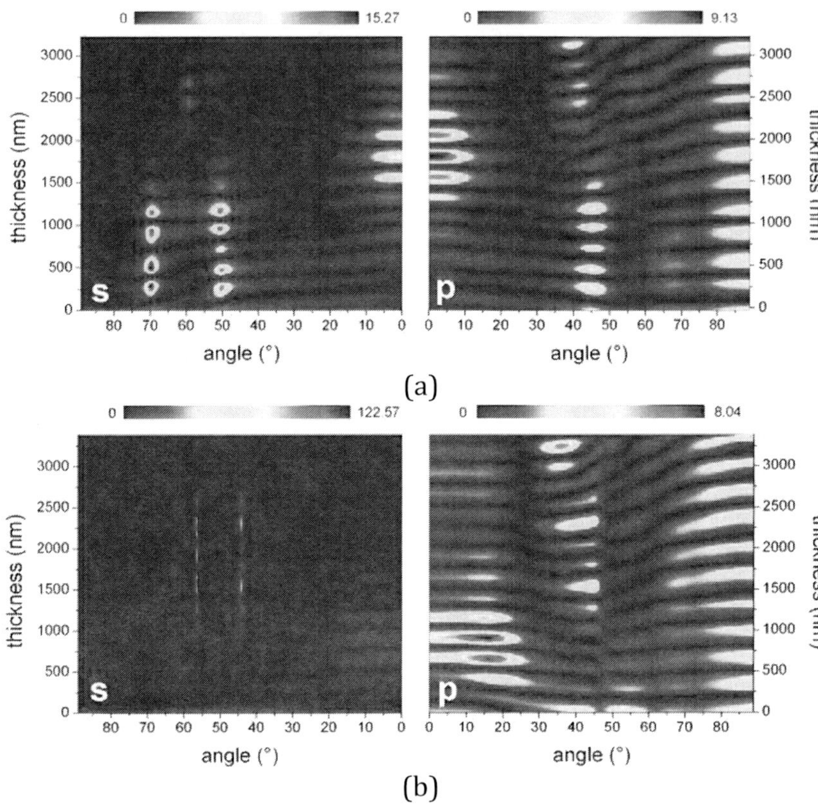

Figure 3. (color online) Dependence of the relative electric field intensity as a function of the incidence angle, for both s and p mode polarizations, in the structures (a) FN4-BM3-FN4 and (b) aBM3-aFN4-aBM3. The particular mode considered corresponds to the photon energies of 1.8955 eV and 1.9335 eV, respectively. These energy positions are indicated by a red circle in the reflectance spectra of each structure in **Figure 2**.

extension of the system for angles of incidence above 50°, whilst for angles between 40° and 50° there is tendency of a larger spatial confinement within the left hand FN4 multilayer; but with no particularly significant values of the intensity (see the corresponding color bar above). In the angular region extending approximately between 20° and 40°, the propagation is very much reduced, with only a small portion of the field penetrating the left FN4 and almost zero intensity in the sandwiched BM3 substructure and the right FN4 portion. The most interesting results regarding the selective spatial localization of the wave field correspond to the interval of oblique incidence with angles between 0 and 10 degrees. Here, the relative field intensity strongly localizes within the BM3 substructure, with the maximum equals to the reported in **Figure 2**(c) for normal incidence; but with the feature that this spatial confinement is kept even for a light incidence with up to 8°.

The s polarization wave for this specific mode shows the same spatial localization within the central BM3 substructure, as compared for p, for angles of incidence between 0 and 8 degrees, approximately. Oblique angles between 8° and 48°, and above 70° for the light incidence lead to almost zero propagation of the s signal in the entire structure. However, there two narrow angular fringes, centered at ~50° and ~68° for which we can observed a strong spatial confinement of the field amplitude within the left FN4 multilayer with separated bright spots having relative intensities above 10, which surpasses the normal incidence highest value in this heterostructure. Again, for angles between those two narrow fringes, the field barely propagates along the hybrid dielectric system.

In the **Figure 3**(b), the results for the dependence of the relative intensity on the angle of oblique incidence correspond to a hybrid aBM3-aFN4-aBM3 structure. In this case, the signal with p-type polarization also displays a rather extended intensity distribution over the entire system for all angles of incidence considered. The maxima of the relative field amplitude confine mostly within the first aBM3 substructure when the angle are between zero and 25° approximately. The highest one appears at about 900 nm from the interface of light incidence and corresponds to a value of 8.04, which is slightly bigger than its value obtained at normal incidence.

The situation for the s-type polarization wave is drastically different in this hybrid heterostructure. As we can readily observe, the propagation of this optical signal throughout such a dielectric multilayer is almost entirely forbidden. The relative field intensity shows very small values in the first half of the structure length, for angles between 0° and 10°. The signal barely travels inside the system for angles larger than 30°, with the exception of two very narrow angular fringes approximately centered at 44° and 57°. For these angles the s-wave propagates along the entire dielectric heterostructure and the relative field intensity is largely localized within the central Fibonacci substructure. It is a significant fact that the oblique incidence of $\sim 57^{\circ}$ boosts the localized intensity corresponding to this field polarization to relative values as high as 122.57.

To complement the discussion above, the **Figure 4** shows the calculated relative intensity maps and the reflectance spectra for the FN4-BM3-FN4 (**Figure 4**(a)) and the aBM3-aFN4-aBM3 (**Figure 4**(b)) hybrid systems. In these cases, a finer angular partition allows to detect the oblique incidence that leads to the largest values of the spatially localized relative s-field intensity. For the hybrid FN4-BM3-FN4 structure, the maximum value is equals to 17.43 and is reached at 69.2 degrees of incident light ray inclination with respect to the first FN4 interface. On the other hand, the value of the maximum $|E|^2$ is obtained, in the case of the aBM3-aFN4-aBM3 heterostructure, when the propagation wavevector of the incident ray forms an angle of 58.2 degrees with respect to the normal direction to the leftmost aBM3 surface (see the diagram of **Figure 1** for reference. Such a value is notably of 137.59. The field intensity in the FN4- BM3-FN4 multilayer shows a selective spatial localization which is mostly confined within the left-hand FN4 substructure. However, the spatial confinement of the optical wave in the aBM3-aFN4-aBM3 structure takes place mainly within the central fourth-generation Fibonacci layered region.

It is worth analyzing the information provided by the calculated reflectance spectra included in Figures 4(a) and 4(b). In the case of the FN4-BM3-FN4 hybrid heterostructure, we can observe a wide photonic band gap for energies above 1.3 eV. Within this gap there are two microcavity propagating modes whose energies are slightly

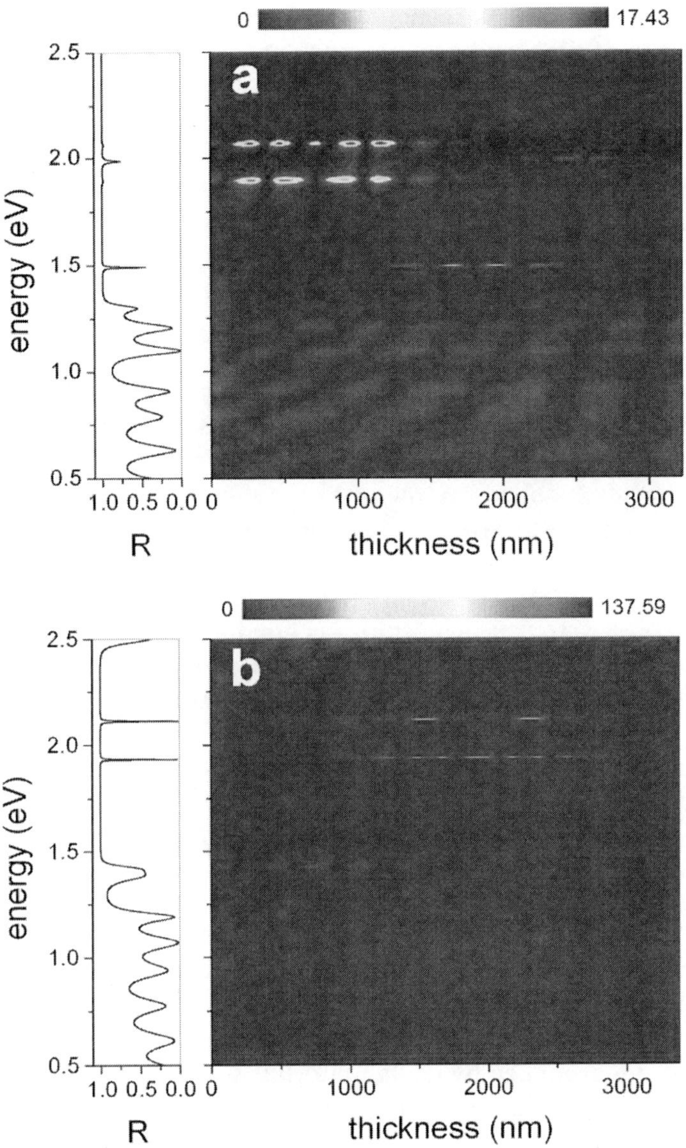

Figure 4. (color online) Enhancement of the confinement of the $|E|^2$intensity for the structures (a) FN4-BM3-FN4 and (b) aBM3-aFN4-aBM3, showed in the **Figure 2**, at the following incidence angles of 69.2° and 58.2°, respectively.

below 1.5 eV and 2.0 eV, respectively. Such a wide gap, obtained for an incidence of 69.2° indicates the prospective usefulness of hybrid heterostructures as omnidirectional mirrors, as it was proposed in reference [51]. The waves that exhibit selective spatial confinement at this particular angle of incidence are not propagating modes but modes with zero transmission. Due to the combined reflecting effects of the Bragg and right-hand Fibonacci multilayers, these waves undergo a process of multiple interferences that leads to the enhancement of the electric field amplitudes within the left part of the structure.

The reflectance spectrum appearing in the Figures 4(b) (case of the aBM3-aFN4-aBM3 hybrid multilayer) also shows a rather wide photonic band gap extending from slightly below 1.5 eV up to a photon energy of approximately 2.4 eV, when the incidence angle is of 58.2°. For this structure, the modes that show selective spatial confinement of the field intensity are the microcavity modes on both sides of the 2.0 eV, which do propagate throughout the entire structure (notice that transmittance is equals to unity for them). The relative field intensity achieves a large value for its maximum, and is confined mostly within the central FN substructure.

Then, according to the information contained in Figures 3 and 4, it can be established that the variation of the angle of incidence of the incoming electromagnetic wave may be used for tuning the presence of a significant selective spatial confinement of the field radiation intensity within hybrid Bragg-Fibonacci dielectric heterostructures.

We must highlight that the results discussed up to this point are related with structures that have a relatively low number of constituent layers (the BM3-FN4-BM3 has 24). It is possible that this number can be reduced by using materials with a different dielectric contrast, and the outcome would be similar. To analyze the influence of the augment in the number of dielectric layers in the system, **Figure 5** contains the calculated relative field intensity for normal incidence in four aBM-aFN-aBM hybrid heterostructures. The order of the central Fibonacci multilayer is kept fixed as fifth. Then, the order of the Bragg-type sandwiching substructures varies from second **Figure 5**(a) to fifth **Figure 5**(d). In the four cases, the highest values of the relative field intensity correspond to signals with amplitudes confined within the FN part. But, as can be readily seen, the maximum amplitudes and their energy positions are quite different from one case to another. For instance, the propagating optical signals that show the greater spatial intensity localization in Figures 5(a) and 5(b) correspond to the three microcavity-like modes associated to the three "A-defects" of the FN5 substructure (AB-AA-BAB-AA-B-AA-B in detail). Whereas, the multilayers with higher orders of the Bragg periodic sequence Figures 5(c) and 5(d) have the spatially localized signal corresponding to a single energy, which is the one corresponding to the lowest of the three confined modes detected in the structures with the lower BM orders. In this sense, for the set of dielectric parameters chosen in the study, the increase in the number of Bragg mirrors makes the entire system to behave as if it were to include a single microcavity-like defect. If we compare the maximum values of the relative field intensities achieved in the case of **Figure 5**(c), it seems like the effects of the three optical defects in the central region merge in such a way that the amplitude of the confined radiation enhances given a maximum value $|E|^2$ equals to ~462. A further augment in one order of the BM generation keeps a stronger localized single-microcavity-like relative intensity, but the maximum value achieved reduces to ~150. What is happening here is that there is a redistribution of the wave field and a substantial part of it confines now within the left-hand BM substructure or simply reflects from the system. A small part of the signal, corresponding to the BM photonic band gap edges, propagates throughout the structure.

The field intensity distribution of the type detected within the higher-order BM substructure in**Figure 5**(d) can be much better observed if we construct a hybrid aFN-aBM5-aFN hybrid multilayer. The **Figure 6** shows the results for the calculated relative electric field intensity as a function of the position in the heterostructure. In this situation, the orders of the Fibonacci sequences used to design the sandwiching FN parts rank from third Figure 6(a) to sixth **Figure 6**(d).

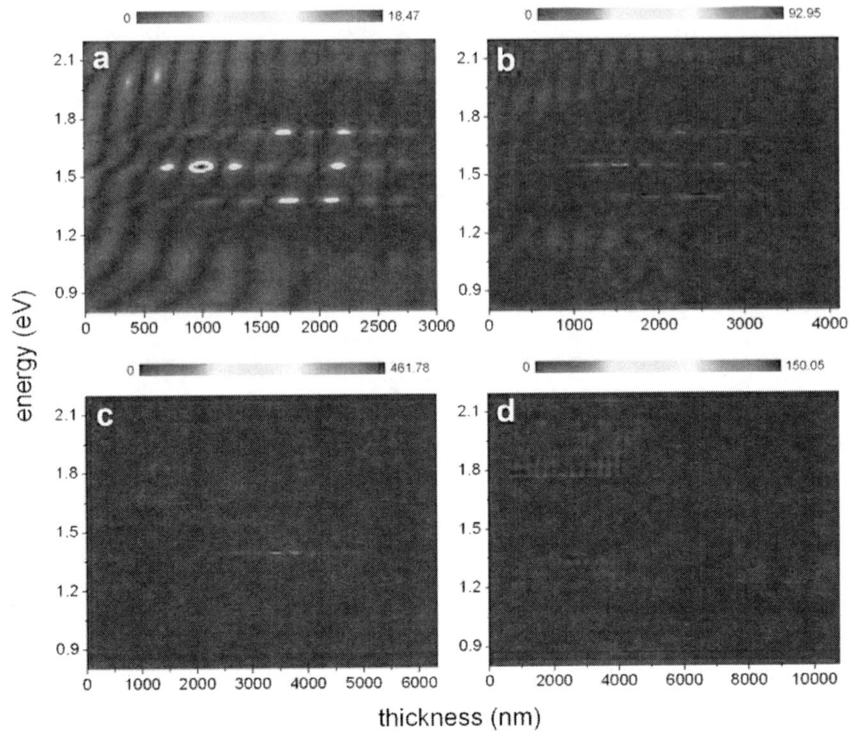

Figure 5. (color online) The relative field intensity of optical modes in a hybrid aBM-aFN-aBM dielectric multilayer as a function of the position in the structure. The order of the central Fibonacci sequence is kept fixed as fifth. The order of the sandwiching BM binary heterolayers varies as following: (a) aBM2; (b) aBM3; (c) aBM4; (d) aBM5.

The general appearance in these four graphics is that of a very well defined BM-related photonic band gap with a direct space picture of allowed and forbidden optical modes of propagation. In the structure possessing the largest FN order **Figure 6**(d), the BM photonic gap remains but the influence of the more disordered FN multilayers destroys the regular arrangement of light intensities above and below the gap displayed in Figures 6(a)-6(c). Also, the largest confined intensities (within the BM part) in this case correspond to a mode with energy above the BM photonic gap (with the higher value of ~112), whilst in the three other cases Figures 6(a)-(c) the maxima correspond to a band gap edge mode.

The BM-related regular distribution of lit and dark fringes in the energy scale is better noticed from the results obtained for the aFN4-aBM5-aFN4 and aFN5- aBM5-aFN5 systems, compared with the hybrid aFN3- aBM5-aFN3. In the cases of FN4 and FN5 components, the allowed modes propagate to a lesser extent within the right-hand FN substructure. Nonetheless, in all the four cases studied, one may observe that the light penetrates largely inside the left-hand FN multilayer, with energies that are within the BM photonic gap. Then, as soon as the waves with those frequencies reach the Bragg central part, they are reflected back.

With the same FN5 sandwiching multilayers, the rise of the order of generation of the central Bragg mirror region leads to the reestablishment of the BM-related field intensity distribution of allowed and forbidden mode intensities discussed above. This can be seen from the **Figure 7**, in which we are presenting the calculated $|E|^2$ for an aFN5-aBM6-aFN5 hybrid heterostructure. In addition, the two profiles of the relative field intensity as a function of the position for the band edge modes are shown. From them, the selective spatial localization of the wave field, as well as the enhancement of the optical intensity are readily seen.

By observing the regular layout of dark and lit frequency regions along the vertical axis of the field intensity distribution in Figures 6 and 7, it is apparent that the design of hybrid FN-BM-FN dielectric heterolayers brings forward the possibility of the practical realization of finite and reduced size photonic crystal systems with a well-defined structure of allowed and forbidden propagation fringes. Besides the selective spatial localization of the wave field intensities, these heterostructures can act simultaneously as optical filters for several intervals of the radiation energy, other than the fundamental gap. We readily notice that the photonic band is an intrinsic feature of the Bragg mirror part of the structure, which is kept

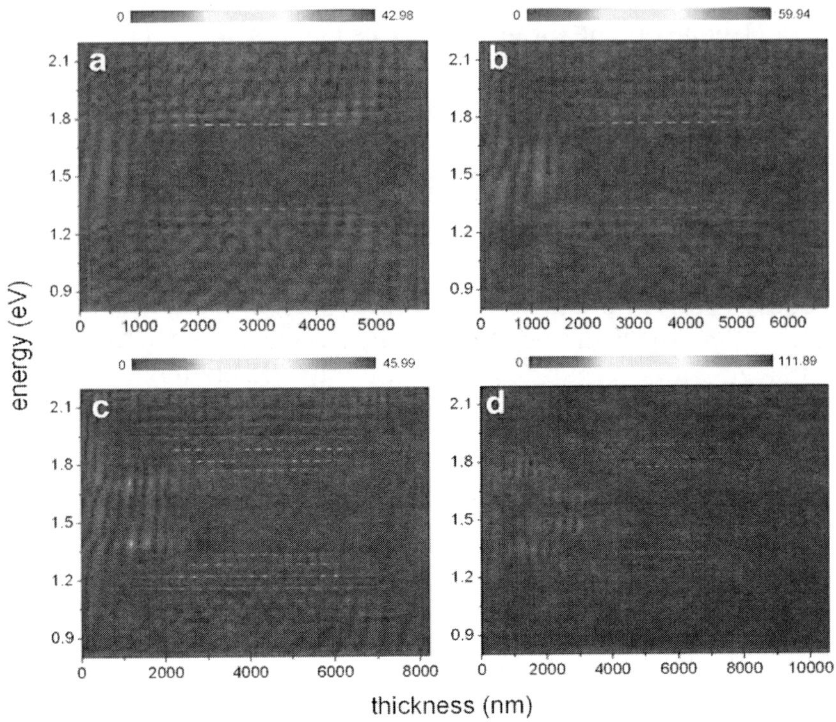

Figure 6. (color online) Distribution of the electric field along the hybrid structures for: (a) aFN3-aBM5-aFN3; (b) aFN4-aBM5-aFN4; (c) aFN5-aBM5-aFN5; (d) aFN6-aBM5-aFN6.

independently of the order of the Fibonacci generation corresponding to the outer substructures. Of course, the better design of these photonic systems would be a compromise between the orders of the periodic and quasiperiodic multilayers, the values of the refractive indices and the layer widths, mainly. Our experience performing numerical simulations with these hybrid structures is, for instance, that the increase of the order of the FN part beyond the fifth in a system like the one used to obtain the **Figure 7** may lead to an important growth in the maximum of the relative field intensity localized within the central BM region. However, the distribution of allowed and forbidden energies practically does not change with respect to that shown in the figure.

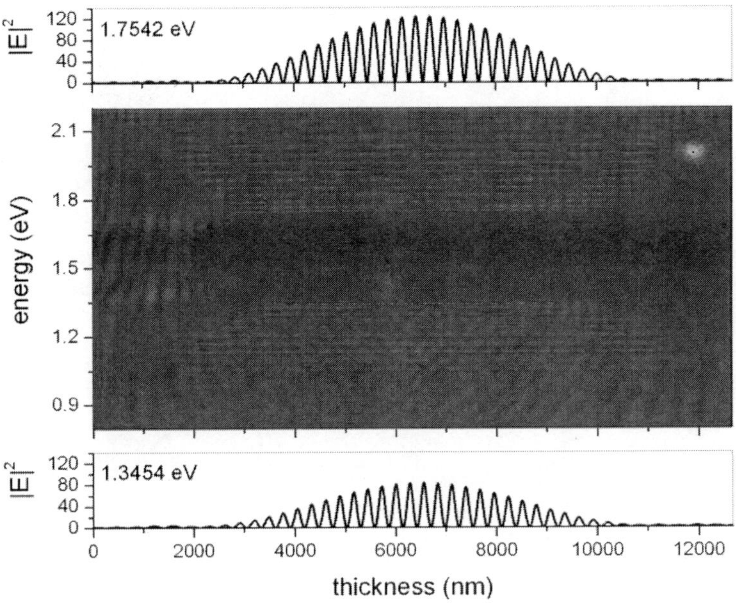

Figure 7. (color online) Distribution of the electric field along the hybrid structures aFN5-aBM6-aFN5 showing a maximum of $|E|^2$ equals to 124.24. The energies of the two particular modes represented correspond to the upper and lower photonic band gap edges.

At this point, it is necessary to briefly comment some possible restrictions to the results presented here. They are associated with the influence of factors that were not included in the model. The calculation performed in this work has assumed ideal conditions such as perfect flat interfaces and no absorption losses. It is clear that interface roughness may lead to reduced reflectance and narrower photonic band gaps given the scattering of the light rays away from the direction of propagation. This was discussed in [20]. In addition, the travel of an optical signal through a real porous silicon-based hybrid heterostructure would suffer from the Rayleigh scattering associated to the intrinsic structural disorder of a porous material, as it was shown in reference [59]. On the other hand, the optical absorption phenomenon would be responsible for a reduction of the optical intensity, thus affecting the values of $|E|^2$.

However, in the case of porous silicon structures, the light absorption takes place mainly for frequencies in the visible blue range and above, whilst at wavelengths around our $\lambda_0 = 800$ nm, the absorption losses are less significant. Despite the possible quantitative modifications that our results might suffer due to these limiting effects, the essence and main qualitative features of the spatial light localization should be kept in a real hybrid structure. Arguments in favor of this conclusion can be found in the satisfactory agreement found in comparisons between theoretical and experimental optical properties of porous silicon quasiregular multilayers published in recent years [32,34,36, 37,39].

4. Conclusions

In this work we have presented a theoretical investigation of the light propagation on hybrid Fibonacci-Periodic dielectric heterostructures. The main purpose of the study was to show the possibility of selective spatial confinement of the optical field intensity within this kind of systems. The results of our calculation reveal that both Periodic-Quasiperiodic-Periodic and Quasiperiodic-Periodic-Quasiperiodic configurations can be used to achieve such a spatial confinement of the light field. In some cases, it is shown that high values of the relative intensity can be obtained in hybrid multilayers made from a quite small number of dielectric layers. Additionally, we have obtained that the variation in the angle of light incidence can be a suitable tool for tuning larger values of the confined field intensity. The calculation has been performed using values of the refractive indices typical of the porous silicon. We believe that the use of constituent materials showing higher dielectric contrast will allow achieve higher values of the confined electric field intensity in hybrid heterostructures. The choice made here for the indices of refraction pursues the possibility of practical realization of this kind of systems via the growth of porous silicon multilayers. That is a cheap procedure by which it is possible the controlled obtention of dielectric layers with n varying from slightly above the unity to 2.5, with layers widths starting from around 7 - 8 tens of nanometers.

Our numerical simulations reveal the possibility of designing multilayer-based photonic crystals with a well defined structure of allowed and forbidden frequencies for the light propagation, using hybrid Fibonacci-BraggFibonacci heterostructures. Given their finite and reduced size, these systems might be of interest for the fabrication of optical devices. In addition, the prediction of the selective spatial localization of some photonic modes, with rather high relative intensity confinement might result of interest in researches looking for plausible lasing regimes using porous silicon dielectric multilayers.

It is worth mentioning that, in order to illustrate the introduction of a hybrid design in the investigation of the spatial mode localization in porous silicon-based onedimensional photonic system, our work has dealt with Bragg-Fibonacci multilayer structures that only involve lower FN or Bragg generations. This is mainly done for the sake of simplicity. Because no previous studies on the subject seem to have been published so far, our goal here was above all to discuss with some extent some physical properties related with the spatial localization of the modes that propagate in the structures. Besides, we choose the simplest configuration, in which “defect-like” layers are made of the same materials that form the layers considered as periodic or quasiregular “mirrors”. In consequence, the results for the spatial localization of the amplitude of the electromagnetic wave can also outcome from a usual “microcavity” theory. However, the use of distinct dielectric configurations in Bragg and quasiregular subsystems (as it was done in [55]), the incorporation of another aperiodic sequences (Thue-Morse, RudinShapiro, etc.), and different geometries (varying the number and widths of the layers) may lead to even more appealing possibilities of application for these heterostructures.

5. Acknowledgements

MEMR thanks hospitality from Escuela de Ingeniería de Antioquia for a sabbatical stay. He acknowledges Mexican CONACYT for support through grants CB-2008 No. 101777, and No. 180636.

REFERENCES

1. E. Yablonovitch, "Inhibited Spontaneous Emission in Solid-State Physics and Electronics," Physical Review Letters Vol. 58, No. 20, 1987, pp. 2059-2062.

2. M. Kohmoto, B. Sutherland and K.Iguchi, "Localization of Optics: Quasiperiodic Media," Physical Review Letters, Vol. 58, No. 23, 1987, pp. 2436-2438.

3. S. John, "Strong Localization of Photons in Certain Disordered Dielectric Superlattices," Physical Review Letters, Vol. 58, No. 23, 1987, pp. 2486-2489.

4. R. Perez-Alvarez and F. Garcia-Moliner, "Some Contemporary Problems in Condensed Matter Physics," Nova Science, New York, 2001, pp. 1-37.

5. R. Perez-Alvarez, F. Garcia-Moliner and V. R. Velasco, "Some Elementary Questions in the Theory of Quasiperiodic Heterostructures," Journal of Physics: Condensed Matter, Vol. 13, No. 15, 2001, p. 3689.

6. J. E. Zarate and V. R. Velasco, "Electronic Properties of Quasiperiodic Heterostructures," Physical Review B, Vol. 65, No. 4, 2001, Article ID: 045304.

7. V. R. Velasco and F. Garcia-Moliner, "Electronic Spectra of Quasi-regular Heterostructures: Simple versus Realistic," Progress in Surface Science, Vol. 74, No. 1-8, 2003, pp. 343-355.

8. E. Maciá and F. Domínguez-Adame, "Electrons, Phonons and Excitons in Low Dimensional Aperiodic Systems," Editorial Complutense, Madrid, 2000.

9. V. R. Velasco, R. Pérez-Alvarez and F. García-Moliner, "Some Properties of the Elastic Waves in Quasiregular Heterostructures," Journal of Physics: Condensed Matter, Vol. 14, No. 24, 2002, p. 5933.

10. H. Aynaou, E. H. El Boudouti, B. Djafari-Rouhani, A. Akjouj and V. R. Velasco, "Propagation and Localization of Acoustic Waves in Fibonacci Phononic Circuits," Journal of Physics: Condensed Matter, Vol. 17, 2005, p. 4245.

11. H. Aynaou, A. Nougaoui, E. H. El Boudouti, D. Bria, V. R. Velasco and B. Djafari-Rouhani, "Comparative Study of the Sagittal Elastic Waves in Metallic and Semiconductor Multilayer Systems between Period and Fibonacci Superlattices," Surface Science, Vol. 584, No. 2-3, 2005, pp. 199-213.

12. C. S. Ryu, G. Y. Oh and M. H. Lee, "Extended and Critical Wave Functions in a Thue-Morse Chain," Physical Review B, Vol. 46, No. 9, 1992, pp. 5162-5168.

13. C. S. Ryu, G. Y. Oh and M. H. Lee, "Electronic Properties of a Tight-Binding and a Kronig-Penney Model of the The-Morse Chain," Physical Review B, Vol. 48, No. 1, 1993, pp. 132-141.

14. V. R. Velasco, J. E. Zárate, "Elastic Waves in Quasiperiodic Structures," Progress in Surface Science, Vol. 67, No. 1-8, 2001, pp. 383-402.

15. J. Tutor and V. R. Velasco, "Some Properties of the Transverse Elastic Waves in Quasiperiodic Structures," International Journal of Modern Physics B, Vol. 15, No. 21, 2001, pp. 2925-2934.

16. S. He and J. D. Maynard, "Detailed Measurements of Ineslatic Scattering in Anderson Localization," Physical Review Letters, Vol. 57, No. 25, 1986, pp. 3171-3174.

17. W. Gellermann, M. Kohmoto, B. Sutherland and P. C. Taylor, "Localization of Light Waves in Fibonacci Dielectric Multilayers," Physical Review Letters, Vol. 72, No. 5, 1994, pp. 633-636.

18. D. Lusk, I. Abdulhalim and F. Placido, "Omnidirectional Reflection from Fibonacci Quasi-periodic One-Dimensional

Photonic Crystal," Optics Communications, Vol. 198, No. 4-6, 2001, pp. 273-279.

19. J. W. Dong, P. Han and H. Z. Wang, "Broad Omnidirectional Reflection Band Forming using the Combination of Fibonacci Quasi-Periodic and Periodic One-Dimensional Photonic Crystals," Chinese Physics Letters, Vol. 20, No. 11, 2003, pp. 1963-1965.

20. V. Agarwal and M. E. Mora-Ramos, "Optical Characterization of Polytype Fibonacci and Thue-Morse Quasiregular Dielectric Structures Made of Porous Silicon Multilayers," Journal of Physics D: Applied Physics, Vol. 40, No. 10, 2007, pp. 3203-3211.

21. N. Liu, "Defect Modes of Stratified Dielectric Media," Physical Review B, Vol. 55, No. 7, 1997, pp. 4097-4100.

22. S. F. Musikhin, V. I. Il'in, O. V. Rabizo, L. G. Bakueva and T. V. Yudinstseva, "Optical Properties of quasiperiodic and aperiodic PbS-CdS superlattices," Semiconductors, Vol. 31, No. 1, 1997, pp. 46-50.

23. R. Pelster, V. Gasparian and G. Nimtz, "Propagation of Plane Waves and of Waveguide Modes in Quasiperiodic Dielectric Heterostructures," Physical Review E, Vol 55, No. 6, 1997, pp. 7645-7655.

24. M. S. Vasconcelos and E. L. Albuquerque, "Transmission Fingerprints in Quasiperiodic Dielectric Multilayers," Physical Review B, Vol. 59, No. 17, 1999, pp. 11128- 11131.

25. C. J. Oton, L. Dal Negro, Z. Gaburro, L. Pavesi, P. J. Johnson, A. Lagendijk and D. S. Wiersma, "Light Propagation in One-Dimensional Porous Silicon Complex Systems," Physica Status Solidi A, Vol 197, No. 1, 2003, pp. 298-302.

26. F. Qiu, R. W. Peng, X. Q. Huang, Y. M. Liu, M. Wang, A. Hu and S. S. Jiang, "Resonant Transmission and Frequency Trifurcation of Light Waves in Thue-Morse Dielectric

Multilayers," Europhysics Letters, Vol. 63, No. 6, 2003, pp. 853-859.

27. L. Dal Negro, M. Stolfi, Y. Yi, J. Michel, X. Duan, L. C. Kimerling, J. Le Blanc and J. Haavisto, "Photon Band Gap Properties and Omnidirectional reflectance in Si/ SiO_2 Thue-Morse Quasicrystals," Applied Physics Letters, Vol. 84, No. 25, 2004, pp. 5186-5188.

28. L. Dal Negro, J. H. Yi, V. Nguyen, Y. Yi, J. Michel and L. C. Kimerling, "Spectrally Enhanced Light Emission from Aperiodic Photonic Structures", Applied Physics Letters, Vol. 86, No. 26, 2005, Article ID: 261905.

29. X. Jiang, Y. Zhang, S. Feng, K. C. Huang, Y. Yi and J. D. Joannopoulos, "Photonic Band Gaps and Localization in the Thue-Morse Structures," Applied Physics Letters, Vol. 86, No. 20, 2005, pp. 201110.

30. S. Chakraborty, D. G. Hasko and R. J. Mears, "Aperiodic Lattices in a High Refractive Index Contrast System for Photonic Bandgap Engineering," Microelectronic Engineering, Vol. 73-74, 2004, pp. 392-396.

31. G. V. Morozov, D. W. L. Sprung and J. Martorell, "Semiclassical Coupled Wave Theory for TM Waves in One-Dimensional Photonic Crystals," Physical Review E, Vol. 70, No. 1, 2004, pp. 016606.

32. M. E. Mora-Ramos, V. Agarwal and J. A. Soto-Urueta, "Propagation of Light in Quasi-Regular Dielectric Heterostructures with Delta-Like Layers," Microelectronics Journal, Vol. 36, No. 3-6, 2005, pp. 413-415.

33. S. Chakraborty, M. C. Parker and R. J. Mears, "A Fourier (k-) Space Design Approach for Controllable Photonic Band and Localization States in Aperiodic Lattices," Photonics and Nanostructures—Fundamentals and Applications, Vol. 3, No. 2-3, 2005, pp. 139-147.

34. V. Agarwal, J. A. Soto-Urueta, D. Becerra and M. E. Mora-Ramos, "Light Propagation in Polytype ThueMorse Structures Made of Porous Silicon," Photonics and Nanostructures—Fundamentals and Applications, Vol. 3, No. 2-3, 2005, pp. 155-161.

35. S. V. Zhukovsky and A. V. Lavrinenko, "Spectral SelfSimilarity in Fractal One-Dimensional Photonic Structures," Photonics and Nanostructures—Fundamentals and Applications, Vol. 3, No. 2-3, 2005, pp. 129-133.

36. V. Agarwal, J. Escorcia-García and M. E. Mora-Ramos, "Optical Properties of Delta Poly-Type Quasiregular Dielectric Structures Made from Porous Silicon," Physica Status Solidi A, Vol. 204, No. 2007, pp. 1367-1371.

37. V. Agarwal, M. E. Mora-Ramos and B. Alvarado-Tenorio, "Optical Properties of Multilayered Period-Doubling and Rudin-Shapiro Porous Silicon Dielectric Heterostructures," Photonics and Nanostructures—Fundamentals and Applications, Vol. 7, No. 2, 2009, pp. 63-68.

38. E. M. Nascimento, F. A. B. F. de Moura and M. L. Lyra, "Suppressed Transmission in Aperiodically Modulated Multilayered Dielectric Structures," Photonics and Nanostructures—Fundamentals and Applications, Vol. 7, No. 2, 2009, pp. 101-107.

39. B. Alvarado-Tenorio, J. Escorcia-García, M. E. MoraRamos and V. Agarwal, "Optical Properties of Non-Periodic Dielectric Systems Made of Nanostructured Porous Silicon," Journal of Nanoparticle Research, Vol. 5, 2009, pp. 69-78.

40. N. Ben Ali and M. Kanzari, "Omni-Directional High Reflectors using One-Dimensional Deformed Quasi-Periodic Cantor Band Gap Structure at Optical Telecommunication Wavelength Band," Mediterranean J. Electron. Commun., Vol. 6, 2010, p. 72.

41. V. Grigoriev and F. Biancalana, "Bistability and Stationary Gap Solitons in Quasiperiodic Photonic Crystals Based on

Thue-Morse Sequence," Photonics and Nanostructures—Fundamentals and Applications, Vol. 8, No. 4, 2010, pp. 285-290.

42. E. Macia, "Optical Engineering with Fibonacci Dielectric Multilayers," Applied Physics Letters, Vol. 73, No. 23, 1998, p. 3330.

43. E. Macia, "Exploiting Quasiperiodic Order in the Design of Optical Devices," Physical Review B, Vol. 63, No. 20, 2001, Article ID: 205421.

44. D. J. Wen, H. Peng and W. H. Zhou, "Broad Omnidirectional Reflection Band Forming using the Combination of Fibonacci Quasi-Periodic and Periodic One-Dimensional Photonic Crystals," Chinese Physics Letters, Vol. 20, No. 11, 2003, p. 1963.

45. A. Montalbán, V. R. Velasco, J. Tutor and F. J. Fernández-Velicia, "Phonon Confinement in One-Dimensional Hybrid Periodic/Quasiregular Structures," Physical Review B, Vol. 70, No. 13, 2004, Article ID: 132301.

46. A. Montalbán, V. R. Velasco, J. Tutor and F. J. Fernández-Velicia, "Selective Spatial Localization of the Atom Displacements in One-Dimensional Hybrid Quasi-Regular (Thue-Morse and Rudin-Shapiro)/Periodic Structures," Surface Science, Vol. 601, No. 12, 2007, pp. 2538-2547.

47. A. Montalbán, V. R. Velasco, J. Tutor and F. J. Fernández-Velicia, "Phonons in Hybrid Fibonacci/Periodic Multilayers," Surface Science, Vol. 603, No. 6, 2009, pp. 938-944.

48. J. Escorcia-García and M. E. Mora-Ramos, "Study on the Influence of the Incidence Direction on the Photonic Band Gap in Porous Si-based Dielectric Heterostructures," PIERS Online, Vol. 5, No. 1, 2009, pp. 36-40.

49. J. Escorcia-García and M. E. Mora-Ramos, "Electromagnetic Modes in Hybrid Periodic-Non-Periodic Dielectric Porous

Silicon Multilayers," PIERS Online, Vol. 5, No. 1, 2009, pp. 91-94.

50. J. Escorcia-García and M. E. Mora-Ramos, "Study of Optical Propagation in Hybrid Periodic/Quasiregular Structures Based on Porous Silicon," PIERS Online, Vol. 5, No. 2, 2009, pp. 167-170.

51. J. Escorcia-García and M. E. Mora-Ramos, "Optical Properties of Hybrid Periodic/Quasiregular Dielectric Multilayers," Superlattices and Microstructures, Vol. 49, No. 3, 2011, pp. 203-208.

52. N. Ben Ali, J. Zaghdoudi, M. Kanzari and R. Kszelewicz, "The Slowing of Light in One-Dimensional Hybrid Periodic and Non-Periodic Photonic Crystals," Journal of Optics, Vol. 12, No. 4, 2010, Article ID: 045402.

53. N. Ben Ali and M. Kanzari, "Designing of Omni-Directional High Reflectors by using One-Dimensional Modified Hybrid Fibonacci/Cantor Band-Gap Structures at Optical Telecommunication Wavelength Band," Journal of Modern Optics, Vol. 57, No. 4, 2010, pp. 287-294.

54. N. Ben Ali and M. Kanzari, "Designing of Stop Band Filters using Hybrid Periodic/Quasi-Periodic One-Dimensional Photonic Crystals in Microwave Domain," Physica Status Solidi A, Vol. 208, No. 1, 2011, pp. 161-171.

55. K. S. Pérez, J. O. Estévez, A. Méndez-Blas, J. Arriaga, G. Palestino and M. E. Mora-Ramos, "Tunable Resonance Transmission Modes in Hybrid Heterostructures Based on Porous Silicon," Nanoscale Research Letters, Vol. 7, 2012, pp. 392.

56. M. E. Mora, R. Perez and Ch. Sommers, "Transfer Matrix in One Dimensional Problems," Journal de Physique Archives, Vol. 46, No. 7, 1985, pp. 1021-1026.

57. P. Yeh, "Optical Waves in Layered Media," John Wiley and Sons, New York, 1988.

58. L. Carretero, M. Perez-Molina, P. Acebal, S. Blaya and A. Fimia, "Matrix Method for the Study of Wave Propagation in One-Dimensional General Media," Optics Express, Vol. 14, No. 23, 2006, pp. 11385-11391.

59. M. Toledo Solano, Y. G. Rubo, J. A. del Río and M. C. Arenas, "Rayleigh Scattering in Multilayered Structures of Porous Silicon," Physica Status Solidi C, Vol. 2, No. 10, 2005, pp. 3544-3547.

5

CHAPTER

Acoustic Emission of Deformation Twinning in Magnesium

Chengyang Mo [1], Brian Wisner [1], Mike Cabal [1], Kavan Hazeli [2], K. T. Ramesh [2], Haitham El Kadiri [3], Talal Al-Samman [4], Konstantin D. Molodov [4], Dmitri A. Molodov [4] and Antonios Kontsos [1,*]

[1]*Theoretical & Applied Mechanics Group, Department of Mechanical Engineering, Drexel University, Philadelphia, PA 19104, USA*

[2]*Hopkins Extreme Material Institute, John Hopkins University, Baltimore, MD 21218, USA*

[3]*Department of Mechanical Engineering, Mississippi State University, Starkville, MS 39762, USA*

[4]*Institute of Physical Metallurgy and Metal Physics, RWTH Aachen University, Aachen 52062, Germany*

Abstract

The Acoustic Emission of deformation twinning in Magnesium is investigated in this article. Single crystal testing with combined full field deformation measurements, as well as polycrystalline testing inside the scanning electron microscope with simultaneous monitoring of texture evolution and twin nucleation were compared to testing at the laboratory scale with respect to recordings of Acoustic Emission activity. Single crystal testing revealed the formation of layered twin boundaries in areas of strain localization which was accompanied by distinct changes in the acoustic data. Testing inside the microscope directly showed twin nucleation, proliferation and growth as well as associated crystallographic reorientations. A post processing approach of the Acoustic Emission activity revealed the existence of a class of signals that appears in a strain range in which twinning is profuse, as validated by the in situ and ex situ microscopy observations. Features extracted from such activity were cross-correlated both with the available mechanical and microscopy data, as well as with the Acoustic Emission activity recorded at the laboratory scale for similarly prepared specimens. The overall approach demonstrates that the method of Acoustic Emission could provide real time volumetric information related to the activation of deformation twinning in Magnesium alloys, in spite of the complexity of the propagation phenomena, the possible activation of several deformation modes and the challenges posed by the sensing approach itself when applied in this type of materials evaluation approach.

Keywords: Acoustic Emission; twinning; Magnesium

1. Introduction

Magnesium (Mg) has a hexagonal close-packed (hcp) structure (c/a=1.624) and a limited number of easy slip systems compared to fcc metals [1]. Basal ⟨a⟩ glide, in addition to ⟨a⟩ prism and pyramidal slip systems, which, however, require higher driving forces and/or elevated temperatures for activation, are known to exist. To accommodate c-axis strains, pyramidal ⟨c+a⟩ slip systems and deformation twinning can occur [2,3,4]. Twins can be both of extension and contraction type, and unlike slip, they generally cause anisotropic changes of the initial crystallographic grain orientation (texture), which depends on the applied forming procedure, the process followed to prepare samples/components and the imposed direction of loading [5]. When a growing twin meets a grain boundary it produces a well-defined tilt that appears as the "needle" structure within the grains. Input load reversibility has been associated with detwinning (i.e., twin annihilation) activity [6,7], which causes additional changes in texture and is related to the observed pseudoelasticity of Mg alloys [8]. Twinning has been further related to redistributions of internal stress/strain that cause: (i) higher order twinning (e.g., daughter twins in grains that had previously formed twins, i.e., parents) [9,10], activation of multiple twin variants [2,11] and double-twinning [4,12]; (ii) activation of "harder" slip systems [8] (e.g., pyramidal ⟨c+a⟩); and (iii) interactions with slip systems that could lead to grain-scale microcracks, which accumulate with loading and ultimately cause fatigue failure [4]. The crystallographic deformation mechanisms in addition to single and polycrystal dislocation-defect interactions in Mg alloys determine their distinct macroscopic mechanical behavior, including asymmetric tension/compression yielding and anisotropic strain hardening [13], which complicate forming procedures and create the need for detailed microstructure-sensitive characterization.

Various methods have been applied to study deformation twinning in Magnesium. Using texture analysis [14,15,16,17,18], $\{10\bar{1}2\}$ twins were observed in single crystal compression [19] while other forms of twins were observed under higher temperature [20]. In addition, heterogeneous extension twins were observed when conducting Channel-Die compression of single crystals [21]. The activation of extension twins and non-basal slips were additionally observed with Transmission Electron Microscope (TEM) by performing nanoindentation and microcompression [22,23]. Electron backscattered diffraction (EBSD) analyses revealed favorable orientations for twinning in Mg polycrystals [24]. Furthermore, twin boundary mobility has been also studied using in situ Scanning Electron microscopy (SEM) observations to calculate twin boundary velocity [25].

Acoustic Emission (AE) is a non-destructive investigation method used, among other applications, in material deformation and damage investigations [13,26,27,28,29,30]. AE activity occurs in all materials under applied loading and can be formally defined as transient elastic waves generated by rapid release of energy caused by events at various length scales, such as crack initiation and propagation, fiber fracture, delaminations, stress corrosion, fatigue and creep. In the case of metals, AE activity has been recorded from dislocation motion/glide [29,30,31,32,33,34], twinning [13,29,35,36], yielding [37], hardening [13] and phase transformations [38]. At the macroscale, the travelling stress waves related to AE are detected even at the small-scale damage level, i.e., long before final macroscopic failure, by specially designed piezoelectric transducers, as well as with a variety of other sensors (e.g., fiber Bragg) and are converted to voltage versus time waveforms. The inverse problem of damage identification based on AE has been greatly enhanced over the past years with the application of mathematical methods, such as neural networks and clustering algorithms that are used to parameterize, identify correlations and statistically group information in pattern recognition algorithms [26,39,40]. Therefore, AE is a diagnostic tool with the potential for the characterization of the mechanical behavior of advanced materials with hierarchical microstructures and multiple phases, in various loading conditions including fatigue, since it can: (i) monitor reversible/irreversible microstructural changes across time and length scales; and (ii) detect dynamic effects, such as dislocation motion and crack initiation.

Various attempts have been made to understand twinning in Mg using AE. Studies have been conducted for AZ [13,41,42], ZK [43] and AM [36] alloys, showing the correspondence between peak of AE count rate and the yield point on stress strain curves. In addition, the effect of grain size and solute content on the AE response was revealed by several studies [44,45,46]. An asymmetry in the AE activity was reported by multiple investigations due to dependence on the loading direction and crystal orientation [36,47]. Furthermore, recent approaches of coupling microstructure and AE have been introduced to support the relationship between AE response and twin nucleation [48,49]. Furthermore, the AE activity during Mg cyclic tests showed the asymmetry of AE response in different loading conditions [34,50,51]. Moreover, machine learning techniques, such as clustering and neural network analyses, have been used to discriminate AE signals from different sources, such as dislocation motions and twin nucleation [52,53]. A similar machine learning approach was used for TRIP/TWIP steels to separate different deformation mechanisms such as stacking faults and martensitic phase transformation [40,54,55]. It should be added here that, apart from work with polycrystalline Mg, only one study is known to the authors that has performed AE monitoring on Mg single crystals [47].

In this article, an approach is proposed to record the AE activity generated by twin nucleation in Mg. To achieve this goal, a comprehensive study of Mg twinning is conducted across length scales and materials, including both single crystal and polycrystals. Digital Image correlation (DIC), scanning electron microscopy (SEM), and electron backscattered diffraction (EBSD) were additionally used to verify the relation of recorded AE signals with activation of deformation twinning. Machine learning techniques were also used to identify characteristics of AE signals generated from twinning. The presented results relate for the first time to the authors best knowledge the AE activity of twin nucleation in single crystals with in situ observed twinning in polycrystals and associated twin-related AE information using laboratory scale specimens.

2. Experimental Procedure

2.1. Mechanical and Nondestructive Evaluation Setup

Three experimental mechanics setups targeting twin activation across different scales coupled with simultaneous recordings of AE activity were used in this investigation. Specifically, Figure 1 shows the first setup consisting of a mechanical stage used for tensile tests of Mg AZ31 polycrystals inside the SEM. The stage is a screw driven Gatan MTEST with a 2000 N load cell which was placed inside an FEI XL30 SEM. A custom built sample grip pre-tilted to 70° was further used for in situ EBSD measurements. The samples were loaded on the stage with a load rate of 0.1 mm/min. EBSD data were collected with EDAX/TSL system at accelerated voltage of 30 kV, spot size 6, aperture size of 100 μm, and working distance of 15 mm with step size of 2 μm. EBSD analysis was performed using EDAX/TSL OIM analysis software. EBSD mapping and SEM images were taken while the samples were held at constant strain.

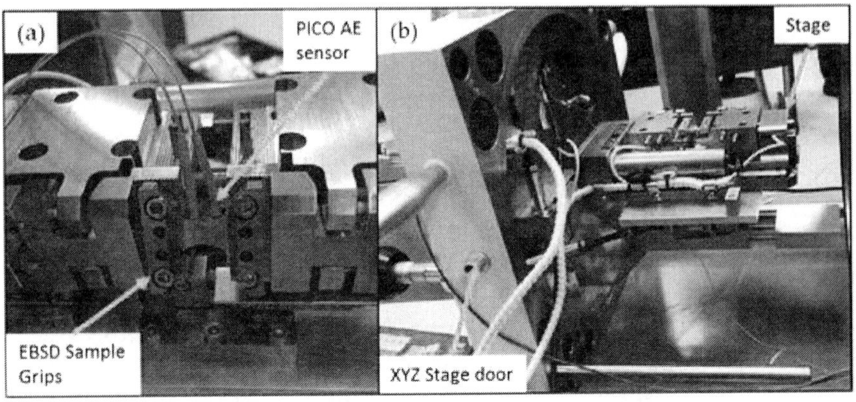

Figure 1. Experimental mechanics setup including a screw-driven stage used for tensile testing of Magnesium polycrystals inside the Scanning Electron Microscope: (**a**) The loading stage with sensors and sample attached on the EBSD grip which is pre-tilted 70°; (**b**) The XYZ stage door with stage on the door.

A MTS universal testing system at room temperature was used for the other two setups including compression tests on Mg single crystals at a constant strain rate of $5 \times 10^{-4} \cdot s^{-1}$ and tension of Mg AZ31 specimens with a constant strain rate of $4.5 \times 10^{-4} \cdot s^{-1}$. Speckle patterns were applied on both the single crystal and polycrystal specimens for strain mapping through the DIC method. 3D DIC results were obtained from a dual camera setup using the commercially available ARAMIS software, as previously shown by the authors in the case of full field deformation measurements of Mg alloys [50,56,57,58].

The AE setup used for all three mechanical testing procedures described above consisted of a AE data acquisition board, a pre-amplifier, and a piezoelectric sensor as described in previous work by the authors [59], while it is worth mentioning that it was the same across all experimental setups, which allows for direct comparisons across scales, specimens, and types of tests. Specifically, AE signals were recorded using PICO piezoelectric sensors, the PCI-2 Acquisition board, and 2/4/6 pre-amplifiers all commercially available by Mistras Group Inc. The AE and DIC systems were linked to load cell voltage readouts which allowed their time synchronization, as shown previously by the authors [60].

2.2. Sample Preparation

Three different types of Mg specimens were tested in this investigation as shown in Figure 2. Magnesium single crystals (Figure 2a,b) were grown in a steel mold under argon atmosphere as described by Molodov [24]. Testing cubes were cut from such single crystals in two orientations for compression testing using a MTS frame. Compression along the c-axis or a-axis was coupled with simultaneous DIC measurements on the corresponding face of the test cubes indicated by the arrows inFigure 1 and AE monitoring using one sensor.

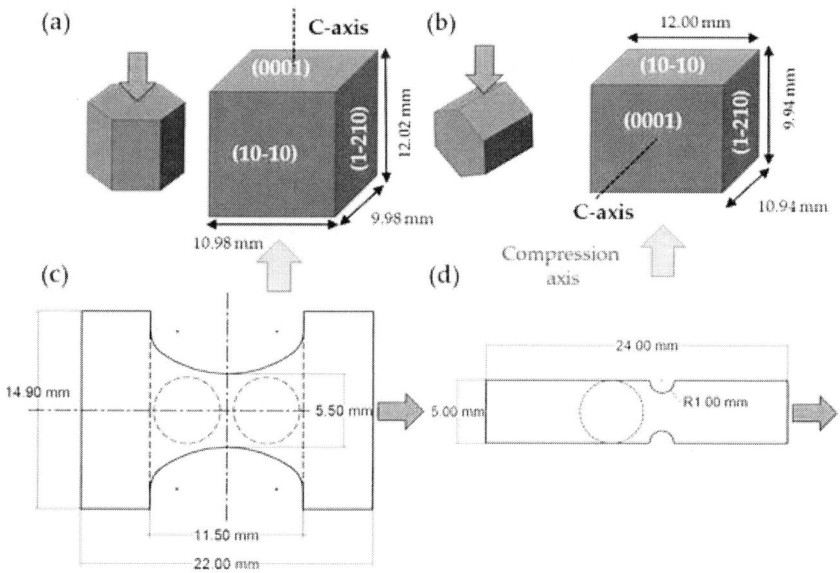

Figure 2. Specimen geometry of: Magnesium single crystal for compression along (**a**) the *c*-axis and (**b**) along the *a*-axis; tensile specimens of Magnesium AZ31 used for (**c**) testing inside the SEM and (**d**) using a MTS frame.

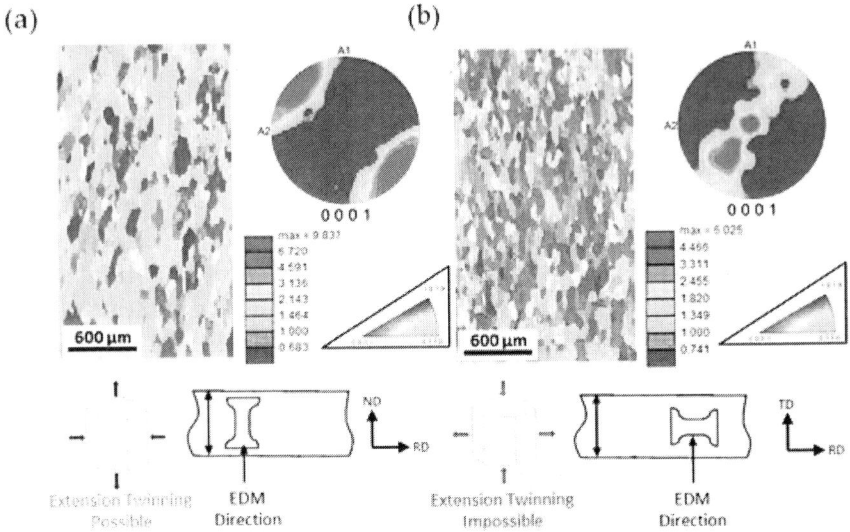

Figure 3. Microstructure of tensile specimens: EBSD map and corresponding pole figures of: (**a**) ND orientation; (**b**) RD orientation.

The tensile samples used for testing inside the SEM chamber (Figure 2c) and with a MTS frame (Figure 2d) were produced from a commercial AZ31 alloy rolled plate, as shown in Figure 3. Cubes were cut out from the 25.4 mm thick plate and were compressed in the Normal Direction (ND) up to 5% compressive strain to increase the dislocation density and stimulate grain growth, as previously described by the authors [50,56]. The cubes were then heat treated at 500 °C for 50 h for grain growth. Larger grains were pursued to lower the activation stress of twinning and to make twinning more observable using the SEM. After heat treatment, samples with a thickness of 2 mm were machined from the cube using Electrical Discharged Machining (EDM). The samples for in situ testing were designed to fit in the stage of Figure 1 with the added capability to attach up to two AE sensors. The samples were ground and then polished with 6 µm, 3 µm, and 1 µm diamond suspension. A mirror finish was achieved with 0.05 µm alumina suspension. The samples were then etched by immersing in a solution of 5% nitric acid, 15% acetic acid, 20%

distilled water, and 60% ethanol for 3 s. The initial microstructures of micro tensile samples are shown in Figure 3. EBSD and texture analysis showed that the RD samples had a predominantly basal texture and therefore were not favorable to deform by extension twinning. ND samples had a prismatic texture and therefore were possible to deform by extension twinning. Both types of samples had an average grain size of 100 μm.

3. Results and Discussion

3.1. Single Crystal Compression

The stress-strain curve combined with the amplitude distribution of the AE signals for an *a*-axis single crystal compression test is shown in Figure 4a. The peak frequency distribution of the AE waveforms whose amplitude distribution is shown in Figure 4a, is plotted Figure 4b. The corresponding full field strains along the loading direction (named *y*-axis for reference) are shown in Figure 4c. It should be noted that the numerical values of strain used in the stress-strain curve of Figure 4a correspond to average of the strain maps shown in Figure 4c. Overall the stress-strain curve appears to have a sigmoidal shape, which is characteristic of twinning activity, marked by a strain plateau that initiates at strain ε_3 and extends up to strain ε_5 [50,57]. This type of load drops agree with previous study [61] that reports load drops during compression of single crystals. This study further suggested that such load drops are possibly caused by dislocation-solute and dislocation-dislocation interactions which can be associated with Lüders band propagation or serrations on the stress-strain curve. In fact, the authors using a continuum plasticity model recently showed that such load drops could indeed be caused by shear band formation [62].

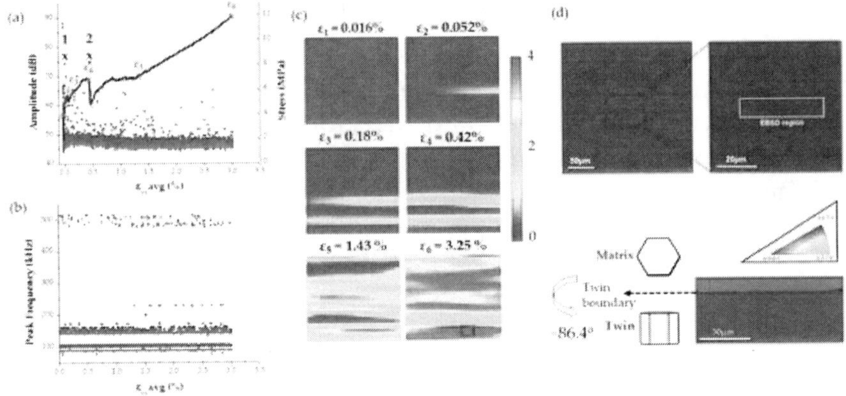

Figure 4. (**a**) Stress-strain curve for a-axis compression specimen coupled with the distribution of the AE waveform amplitudes; (**b**) corresponding distribution of AE signal peak frequency values; (**c**) full field strain maps along the loading direction and (**d**) region definition for EBSD measurements and schematic that explains the formation of a twin boundary.

Most of the AE signals in this test were found to have amplitude values between 40 and 50 dB. However, some higher amplitude (>65 dB) AE signals were observed at strain value $\varepsilon_2 = 0.052\%$ at which point the stress-strain curve appears to show a small load drop followed by a slope change. Interestingly, at this strain value a horizontal strain localization appears as shown in Figure 4c. This strain localization intensifies and a new localization band parallel to the one observed at ε_2 appears at strain ε_3, followed by a more significant now load drop at strain value ε_4,which is accompanied by a second jump of the AE amplitude distribution, as shown in Figure 4a. Past strain ε_4 the stress increases again non-linearly with respect to the average strain along the loading direction, while at strain ε_5(=1.43%) the slope of this curve increases. Past strain ε_5 the full field strain maps appear to have multiple localization zones until failure at strain ε_6.

Strain localization bands during monotonic and cyclic loading of Mg polycrystals were first reported by the authors [50,56,57,58] and others [63,64], while their appearance was spatially correlated with the inhomogeneous activation of twinning [57]. Based on such prior information and since to the authors best knowledge this is the first time that such results are reported for single crystals of Mg, a region was selected based on the electron microscopy images shown in Figure 4d to perform EBSD measurements. The EBSD scan revealed the presence of extension twinning as confirmed by the measured misorientation angle of 86.4°. As a result, there is evidence to suggest that the strain localizations seen in the DIC strain maps correspond to twinning activity which in this case appears to create a horizontal twin boundary as explained schematically in Figure 4d.

Two extracted features of the recorded AE signals that are identified in Figure 4 include the distinct high amplitude values at given time/strain instances during this monotonic compression test, as well as the appearance of relatively high (~500 kHz) peak frequency values, which both agree with reports found in previous investigations [48,65]. The coupling of such AE data with full field strains allows the direct correlation of such features with the initiation of a clearly defined strain localization band, as seen in Figure 4c. Moreover, load drops are observed exactly at loading increments were the high amplitude AE waveforms were recorded. This observation agrees with a previous investigation [61] that reports load drops after twin nucleation. Hence, AE signals, such as the ones shown in Figure 5, which were recorded in instances when bulk load drops, are very possibly directly related to initiation of twinning in Mg single crystals. These AE signals consist of burst type waveforms with distinct peak frequency at ~500 kHz confirmed by wavelet transform, which also corresponds to the sharp increase of amplitude that occurs at the time window defined by these waveforms.

To compare the results shown in Figure 4 with *c*-axis compression data, Figure 6 shows the stress–strain curve coupled with AE amplitude distribution and corresponding full field strain data. In this case, the stress-strain curve is a smoothly nonlinear typical of a ductile metal, while the AE response shows no outliers with all related signals having amplitude between 50 and 65 dB. In addition, no strain localization in the DIC strain maps is observed in this case until near the end of the experiment when high strains were observed near the edges of the specimens coinciding with the locations of the compression platens used to apply the load. Two randomly-selected AE waveforms from this experiment are shown in Figure 6d, which shows that most AE signals from *c*-axis compression are of continuous type and hence different compared to the burst type signals recorded during a-axis compression and shown in Figure 5.

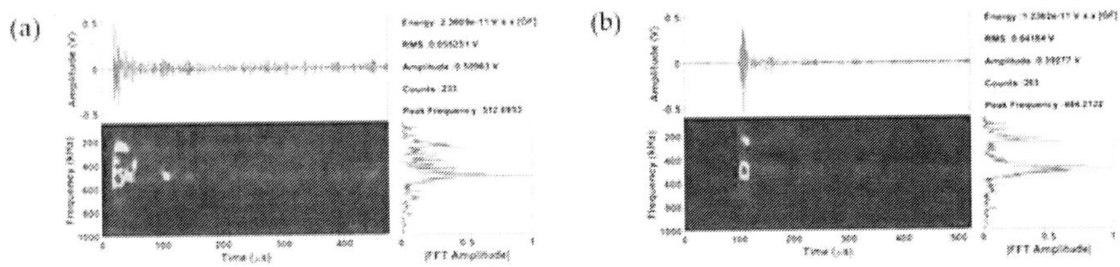

Figure 5. Two identified AE signals related to initiation of twinning in *a*-axis compression of Mg single crystals: (**a**) Signal 1 identified inFigure 4a; (**b**) Signal 2 identified in Figure 4a.

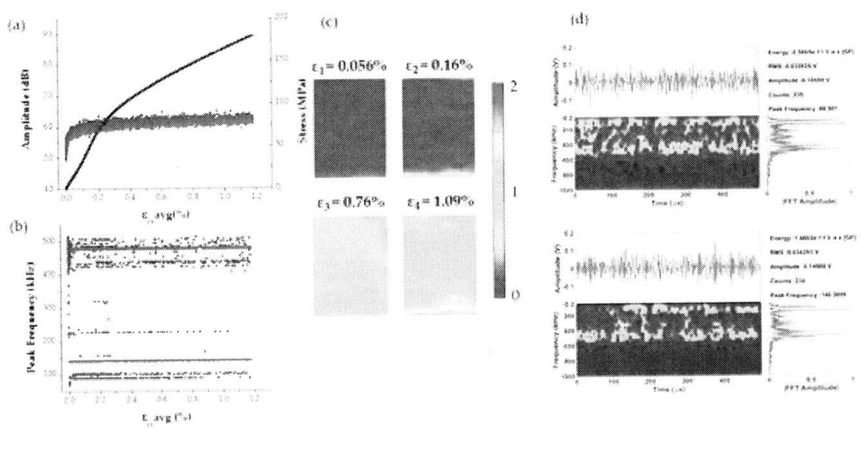

Figure 6. (**a**) Stress-strain curve for c-axis compression specimen coupled with the distribution of the AE waveform amplitudes; (**b**) corresponding distribution of AE signal peak frequency values; (**c**) full field strain maps along the loading direction; (**d**) randomly selected AE signals.

3.2. AZ31 Tensile Tests Inside the SEM

The corresponding stress-strain curve coupled with the amplitude distribution of AE signals in the case of a tensile test of AZ31 specimens cut along both the ND and RD directions are shown in Figure 7. Similar to the single crystal tests, anisotropic mechanical behavior is observed for the two types of samples. The ND sample has a yield stress of ~50 MPa, while the RD sample has a yield stress of ~60 MPa. Due to the geometry of the specimens, slippage was observed at the grips when high load were reached and therefore the specimens did not fail. High amplitude AE signals were only observed near yield points for both samples which have similar maximum amplitudes. It should be further noted that in order to monitor the evolution of texture and the initiation of twinning, DIC measurements were not performed in these experiments and, therefore, the strain values in Figure 7 correspond to engineering strain computed given the displacement readings by the stage.

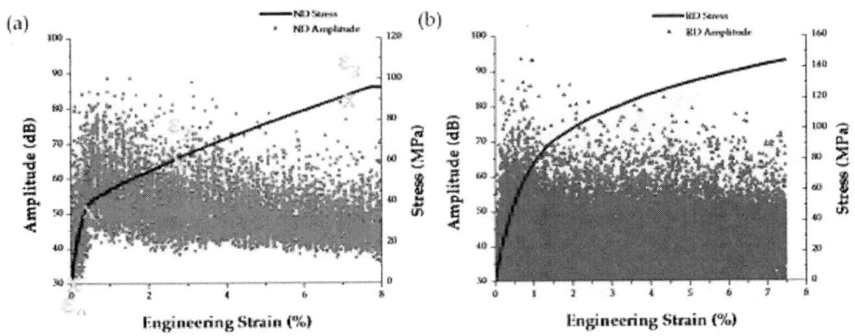

Figure 7. Stress strain curve and corresponding AE amplitude distribution of tensile tests of: (**a**) ND and (**b**) RD specimens.

To compare the AE response in Figure 7 in these two tests, direct microstructure monitoring using EBSD measurements and SEM images is shown in Figure 8. The ND specimen shows severe twinning activity while the RD samples does not. Specifically, in situ obtained EBSD maps at the indicated strain values on the stress-strain curve of the ND sample are shown in Figure 8a. It can be clearly seen that the original microstructure shows a relatively twin free texture with some remnant twins due to the thermomechanical processing. At strain value ε_1 twins appear in several grains. Further twin nucleation and growth are observed in the previously twinned grains at ε_2. At ε_3, some grains are fully reoriented to basal texture. SEM images of the same area before and after the test are shown in Figure 8b. Surface extrusion and intrusion after twin nucleation are observed, as described previously by the authors [56]. The distribution of misorientation angle at ε_0 and ε_2 is shown in Figure 8c, from which extension twinning can be confirmed by the fact that several occurrences of misorientation angle between 85 and 90 degrees is found. The misorientation angle is calculated between identified grains with grain tolerance angle of 5 degrees.

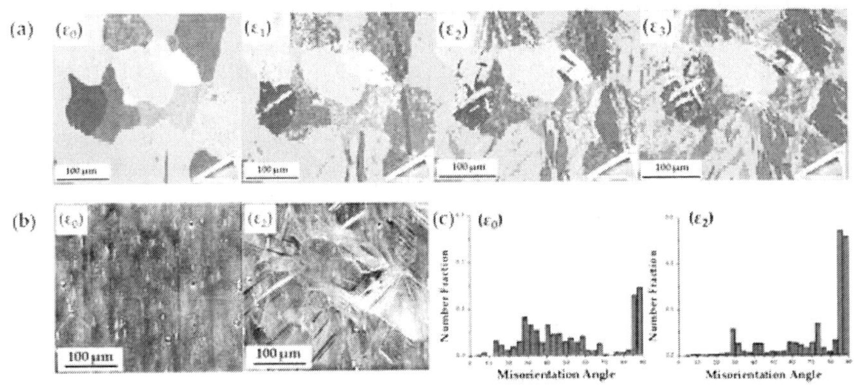

Figure 8. (**a**) Texture evolution and twin activity monitoring of the ND specimen at corresponding strain values marked in the stress-strain curve of Figure 7; (**b**) SEM images at corresponding strain values ε_0 and ε_2; (**c**) misorientation angle distribution between 0 and 90 degrees at corresponding strain values ε_0 and ε_2.

Further analysis of the AE results shown in Figure 7 was performed to identify specific information related to the twinning activity. Specifically, a comparison of the peak frequency values shown in Figure 9 indicates that the ND specimen has more AE signals with dominant frequency in the 450–500 kHz range and less signals in the 100–200 kHz. This is expected, as Li [65] also showed that a twinning AE signal will have a peak frequency around 490 kHz. Note that the AE signals in the single crystal tests previously reported in Figure 5 also have such a peak frequency in this range, which provides additional evidence that this frequency-based AE feature is useful to flag the initiation of twinning activity.

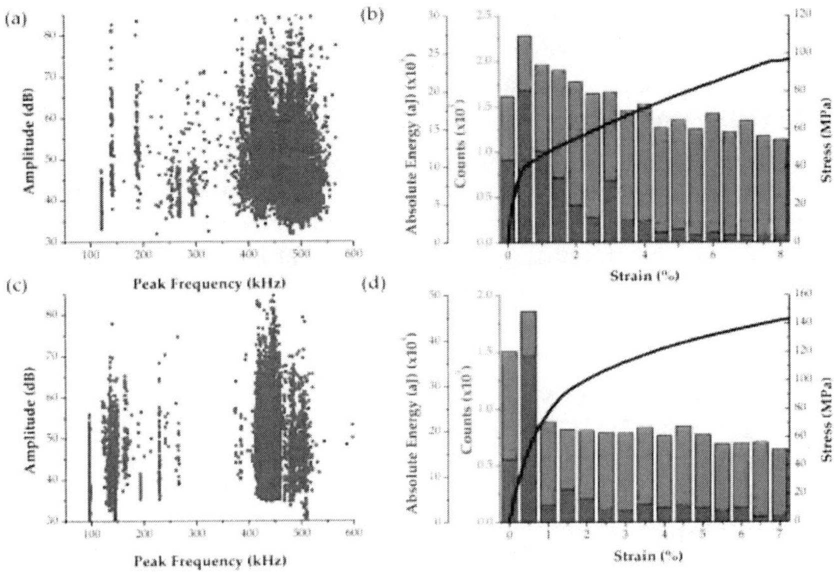

Figure 9. Analysis of the AE activity of the specimens tested inside the SEM: (**a**,**c**) amplitude vs. peak frequency plot and; (**b**,**d**) distribution of AE counts and absolute energy for ND (**top**) and RD (**bottom**) specimens.

Distributions of AE counts and absolute energy of AE signals as a function of engineering strain are shown in Figure 9b,d. The AE counts and absolute energy distributions peak near the yield point. It has been shown that AE responses are most intensive near yield point in Magnesium due to twin nucleation and massive dislocation multiplication [41,42,66]. In the ND direction, twinning should be the major source of AE signal during the early stages of deformation, as also supported by the direct microstructure results shown in Figure 8. In contrast, in RD samples dislocation motion should be the main source of AE signal as extension twins are not possible for this crystal orientation. In addition, it can be further observed that the AE counts and absolute energy decrease after yielding. However, such a drop is much more rapid in the RD compared to the ND specimen. This observation could be explained by the fact that, in the ND specimen, twin nucleation should be prominent near the macroscopically defined yield point, which explains the peak in both counts and absolute energy for burst type emissions such as the ones reported in Figure 5. The slow drop in counts and absolute energy past yielding could be attributed, based also on the in situ obtained results in Figure 8, to the fact that twins are still nucleating, growing and interacting after the yield point. Consequently, the counts and absolute energy distributions present a rather slow decrease until ~3%, which previous work by the authors has shown that it corresponds to the end of the plateau and the onset of hardening in the stress strain curve of specimens prone to twinning [57].

To further classify differences in the AE data for the specimen tested inside the SEM, a clustering approach was followed based on the use of the K-means method with the following features of amplitude, AE counts, absolute energy, and peak frequency according to prior work of the authors [26,28] and based on the AE signals shown in Figure 7 and Figure 9. Clustering analysis is a pattern recognition technique which groups data according to their similarity. Eight classes were identified for signals with peak frequency between 350 and 550 kHz which is the frequency range in which both single crystal and SEM tests have showed that twinning AE signals exist, as shown in Figure 10a. The optimum number of classes was selected using three performance indices, namely the R, τ, and S values, where R is the Davies-Bouldin index, τ is the minimum cluster center distance divided by maximum intercluster distance as defined by Tou [67], while S is the Sihouette index. In general, low values for $R$and high value of τ and S are desired for the clustering to be well-defined. The computed R, τ, and Svalues as a function number of classes are plotted in Figure 10b. Even though a three class clustering appears to have the highest τ and S and lowest R value, it does not produce a class of AE signals that were only existent in the ND direction which was desirable. On the other hand, the eight class clustering shown in Figure 10a, has a local minimum in R and a local maximum for τ as shown in Figure 10b. Therefore, it could be argued that clustering with eight classes not only produces an acceptable classification based on the three criteria used in this investigation, but it also leads to two classes, class 5 and class 7, which exist only in the AE dataset of the ND specimen, as shown by plotting the AE data in their projection in the corresponding stress–strain curves shown in Figure 10c–g. Note that such projection was possible, due to the synchronization between AE, DIC and load data which allows the so called "hit-driven" projection of AE data using two external to the AE system parameters, in this case stress and strain values. A comparison between class 5 and class 7 data shows that AE signals in class 5 generally have lower amplitudes compared to class 7 data. Although the authors are well aware that amplitude is not a reliable classifier in such analyses, the high amplitude of the signals in class 7 combined with their larger number of counts (as shown in Figure 10a), the results in Figure 9 (which confirms large number of counts for twinning AE activity) and both the single crystal (Figure 4) and SEM test data (Figure 7) previously reported, lead to the possibility of these signals to be representative of AE information from twinning.

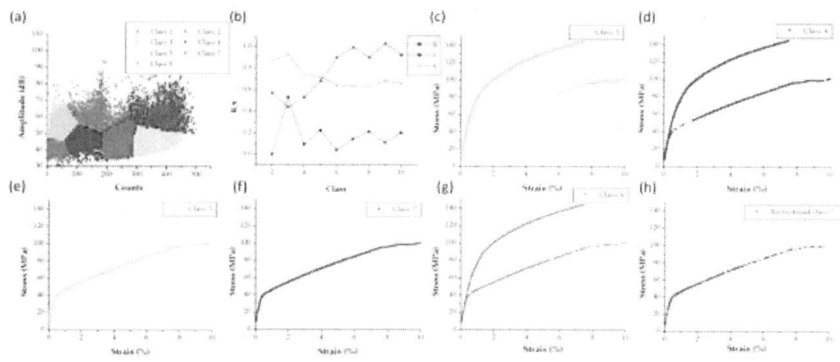

Figure 10. (**a**) K-Mean clustering results based on the combined AE datasets of both ND and RD tests that have a Peak Frequency between 350 and 550 kHz; (**b**) computed R, τ and S values against number of classes; (**c–g**) class 3–5 and 7–8 AE hits plotted as a projection on the ND and RD stress strain curves; (**h**) reclustered class 7 AE hits.

Further analysis of all AE waveforms within class 7 revealed both continuous and burst type emissions. To further classify the burst type signals, which are more possible to originate from twinning activity, as shown in Figure 5, a new feature extraction with adaptive threshold was performed on all AE signals within this class. Subsequently the data were clustered again using the features of duration and risetime, motivated by the fact that burst type signals by definition have short duration and risetime. The reclustered AE hits are projected on the ND load curve shown in Figure 10h. In addition, the amplitude vs. strain of the reclustered class 7 signals is shown in Figure 11a. Interestingly, this projection of the clustered data shows that most of the AE signals in this class occur during early stages of the test (<3%). Moreover, Figure 11b,c show that AE signals in class 7 have much larger number of counts and the highest absolute energy compared to all other classes in the strain range that in Figure 8was directly associated with profuse twinning. In addition, the cumulative absolute energy of this reclustered dataset appears to increase drastically near the yield point. This is consistent with the results in Figure 9b, which suggested that signals with the most energy can be found near the yield point. The AE information of the signals within the reclustered class 7 is, therefore, both statistically and physically meaningful to be correlated with twinning activity in Mg. To further evaluate such information, the AE activity of tension tests using a standard MTS frame for the specimens shown in Figure 2c was further analyzed, as shown in the following figures.

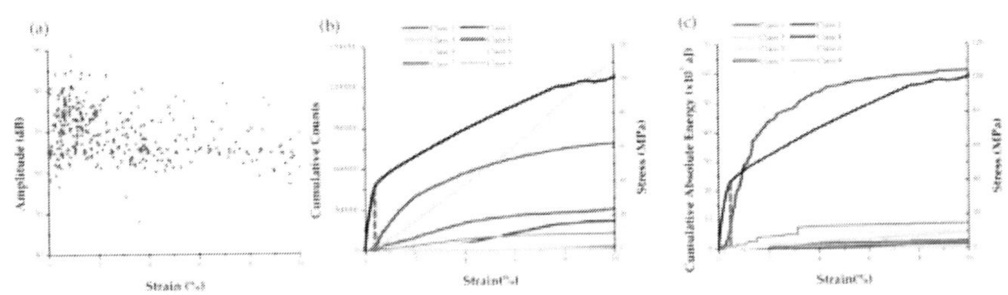

Figure 11. (**a**) Amplitude vs. strain of the reclustered class 7; (**b**,**c**) cumulative AE counts and absolute energy of all classes as a function of strain and compared with the stress–strain curve of the ND specimen.

3.3. AZ 31 Tensile Tests Using the MTS

The procedure developed for the tensile tests inside the SEM was further applied to the MTS level tests using the specimens shown in Figure 2c. The authors have recently shown that this type of geometry due to the strain localization zone between the two notches twins profusely and therefore the AE datasets obtained could be used for comparisons with the tests inside the SEM [68]. The overall mechanical behavior for both ND and RD specimens combined with the AE activity is shown in Figure 12a, while Figure 12b shows the peak frequency of this AE activity. In this case amplitude, absolute energy, and peak frequency were chosen as the descriptors used in clustering. Again, eight classes were determined using the same procedure as described in the previous section and only these AE signals shown in Figure 12b with a peak frequency in the 400–600 kHz range. The clustering results are shown in Figure 12c–g. Using the same criteria described in the previous section, class 8 was identified as the twinning class and therefore it was reclustered to select only the burst type signals.

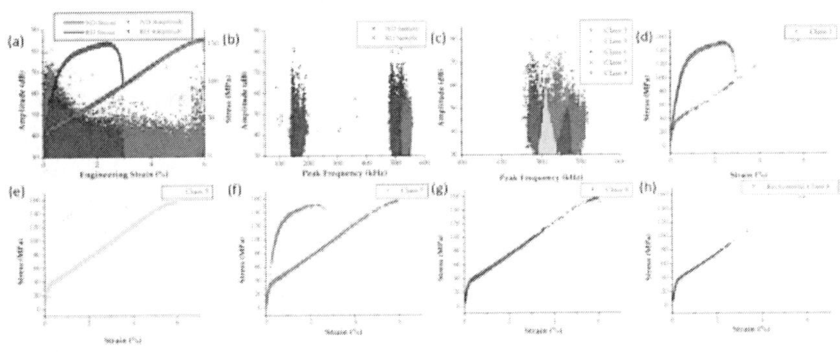

Figure 12. (**a**) Stress-strain curves and amplitude plot of ND and RD test samples at the MTS level; (**b**) frequency amplitude plot of both samples; (**c**) K-Means class distribution of data with peak frequency between 400 and 600 kHz; (**d**–**g**): class 3, 5, 7, 8 AE hits; (**h**) reclustered class 8 AE hits.

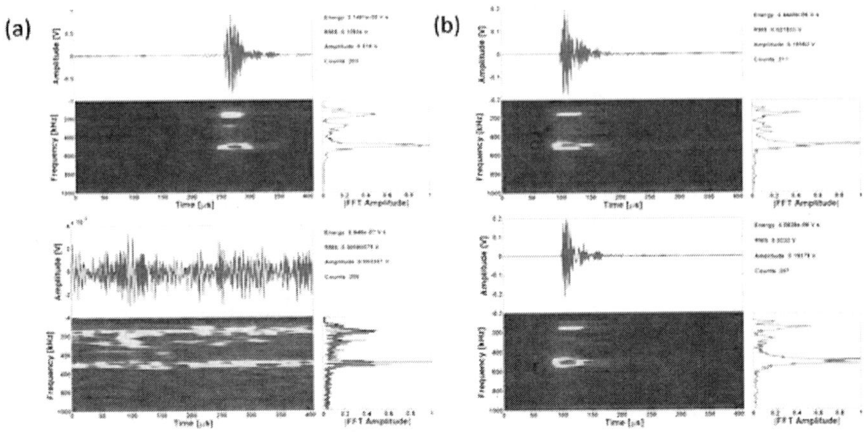

Figure 13. (**a**) Randomly selected signals from class 8 in Figure 12showing both continuous and burst type emissions; (**b**) randomly selected signals from reclustered class 8 containing only burst emissions.

Some representative waveforms of the original class 8 are shown in Figure 13, which contains both continuous and burst type waveforms. However, the reclustering procedure showed that only burst type signals can be found in the new class 8, similar to the procedure followed for the datasets obtained from testing inside the SEM.

3.4. Summary of AE Signals by Twinning in Magnesium

Based on the results presented in this section, a summary of AE activity characteristics associated with twinning is shown in Table 1. The results based on the clustering from the polycrystalline specimens (both inside the SEM and at the MTS level) are found to agree with the results obtained by single crystal testing (twin 1 and twin 2 referring to the highlighted AE hits in Figure 4a). Specifically, the AE signals identified as twin related were of burst nature with relative high (~500 kHz) peak frequency and frequency centroid values, as well as relatively high rise and decay angles. The high rise and decay angle of AE signals captured in this study agree with a previous report [13], as nucleation of twins produce burst type AE signals. The high frequency of twinning AE signals has also been previously reported [46,53,65]. Regarding their amplitude, it makes physical sense that the amplitude of AE signals in the single crystal tests is higher than the polycrystals, as the twins produced in the single crystal testing are in general larger and are activated easier since there is no obstacles (e.g., grain boundaries) or other effects (e.g., texture) which would affect the activation of this deformation mechanism. In contrast, twin activation in polycrystals is highly dependent on crystallographic orientation and grain size as previously shown by the authors [56,57].

Table 1. Comparison of AE features related to twinning activity in Magnesium.

AE Features of Twinning	Single Crystal Twin 1	Single Crystal Twin 2	SEM AZ31 Twin Clustering (388 hits)	MTS AZ31 Twin Clustering (410 hits)
Amplitude (dB)	74	75	69 ± 6	60 ± 8
Rise Angle	0.721	0.724	0.92 ± 0.41	0.51 ± 0.45
Decay Angle	0.838	0.913	0.89 ± 0.38	0.46 ± 0.42
Peak Frequency (kHz)	513	484	465 ± 37	507 ± 14
Frequency Centroid (kHz)	471	500	502 ± 70	520 ± 68

The quantitative characteristics of AE signals from twinning are in agreement with previous studies [13,36,40,41,42,43,44,52,53,65,66]. To the best of the authors' knowledge, no such characteristics of AE signals from twining have been reported directly. Furthermore, unlike previous machine learning attempts using adaptive sequential K-means method [53], this study uses a two-step K-means clustering to obtain those AE signals that are most probable to be associated with twinning. Such results further benefit from comparisons between Magnesium single crystal and polycrystal testing, which deviates from the AE analysis done previously [47].

4. Conclusions

The results in this article relate twinning activity across scales with Acoustic Emission information. To the best of the authors' knowledge, this is the first time that testing of Magnesium single crystals is compared with both testing inside the scanning electron microscope and at the standard laboratory scale using the same type of Acoustic Emission monitoring system and paired, when possible, with full field deformation measurements. The mechanical test data presented in this article, directly show that twin activation at both single crystal and polycrystalline level is associated with distinct AE activity, which was cross-validated by the overall stress strain curve (by seeing load drops and stress plateau regions formed), as well as microscopy information. Such microscopy revealed twin boundary formation at locations where strain localizations appear in single crystal testing, as well as texture and in situ twin nucleation and growth in polycrystals tested inside the electron microscope. The availability of these datasets and the fact that such AE activity could be correlated directly with both mechanical test data and microscopy led to the application of a post processing method, which resulted to the definition of ranges for features of Acoustic Emission signals which appear to be related to twinning. This approach, provides evidence of the applicability and usefulness of nondestructive evaluation based on acoustics to assist characterization and monitoring of dominant deformation mechanisms in advanced materials.

Acknowledgments

The corresponding author would like to thank National Science Foundation (NSF) for the financial support provided through the CMMI #1434506 award to Drexel University. He also acknowledges the technical support received under the National Aeronautics and Space Administration Space Act Agreement, No. SAA1-19439 with Langley Research Center. Three of the authors (Talal Al-Samman, Konstantin D. Molodov and Dmitri A. Molodov) express their gratitude to the Deutsche Forschungsgemeinschaft (DFG) for financial support (Grants AL 1343/5-1 and MO 848/18-1). Furthermore, Sean R. Agnew from University of Virginia is acknowledged for providing useful comments on the results of this article.

Author Contributions

Chengyang Mo, Brian Wisner, Mike Cabal and Antonios Kontsos prepared the specimens and performed all mechanical tests. They further made the Digital Image Correlation and Acoustic Emission measurements and analysis shown in this manuscript. In addition, they performed the microscopy on the specimens tested inside the microscope. Kavan Hazeli and K. T. Ramesh assisted with testing, Digital Image Correlation and microscopy of the single crystal specimens. Haitham El Kadiri, Talal Al-Samman, Konstantin D. Molodov and Dmitri A. Molodov prepared the single crystal specimens and assisted with the discussion presented in this article.

Conflicts of Interest

The authors declare no conflict of interest.

References

1. Hull, D. *Introduction to Dislocations*; Pergamon Press Ltd.: Oxford, UK, 1965.

2. Yoo, M.H. Slip, twinning, and fracture in hexagonal close-packed metals. *Metall. Trans. A* **1981**,*12A*, 409–418.

3. Kalidindi, S.; Salem, A.; Doherty, R.D. Role of deformation twinning on strain hardening in cubic and hexagonal polycrystalline metals. *Adv. Eng. Mater.* **2003**, *5*, 229–232.

4. Ando, D.; Koike, J.; Sutou, Y. Relationship between deformation twinning and surface step formation in AZ31 magnesium alloys. *Acta Mater.* **2010**, *58*, 4316–4324.

5. Knezevic, M.; Levinson, A.; Harris, R.; Mishra, R.K.; Doherty, R.D.; Kalidindi, S.R. Deformation twinning in AZ31: Influence on strain hardening and texture evolution. *Acta Mater.* **2010**, *58*, 6230–6242.

6. Matsuzuki, M.; Horibe, M. Analysis of fatigue damage process in magnesium alloy AZ31. *Mater. Sci. Eng. A* **2009**, *504*, 169–174.

7. Wu, L.; Agnew, S.R.; Ren, Y.; Brown, D.W.; Clausen, B.; Stoica, G.M.; Wenk, H.R.; Liaw, P.K. The effects of texture and extension twinning on the low-cycle fatigue behavior of a rolled magnesium alloy AZ31B. *Mater. Sci. Eng. A* **2010**, *527*, 7057–7067.

8. Wu, L.; Agnew, S.; Brown, D.; Stoica, G.; Clausen, B.; Jain, A.; Fielden, D.; Liaw, P. Internal stress relaxation and load redistribution during the twinning–detwinning-dominated cyclic deformation of a wrought magnesium alloy, ZK60A. *Acta Mater.* **2008**, *56*, 3699–3707.

9. Barnett, M.R.; Keshavarz, Z.; Beer, A.G.; Ma, X. Non-Schmid behaviour during secondary twinning in a polycrystalline magnesium alloy. *Acta Mater.* **2008**, *56*, 5–15.

10. Molodov, K.D.; Al-Samman, T.; Molodov, D.A.; Gottstein, G. Mechanisms of exceptional ductility of magnesium single crystal during deformation at room temperature: Multiple twinning and dynamic recrystallization. *Acta Mater.* **2014**, *76*, 314–330.

11. El Kadiri, H.; Kapil, J.; Oppedal, A.L.; Hector, L.G.; Agnew, S.R.; Cherkaoui, M.; Vogel, S.C. The effect of twin–twin interactions on the nucleation and propagation {10–12} of twinning in magnesium. *Acta Mater.* **2013**, *61*, 3549–3563.

12. Beyerlein, I.J.; Wang, J.; Barnett, M.R.; Tome, C.N. Double twinning mechanisms in magnesium alloys via dissociation of lattice dislocations. *Proc. R. Soc. A* **2012**, *468*, 1496–1520.

13. Lou, X.; Li, M.; Boger, R.; Agnew, S.; Wagoner, R. Hardening evolution of AZ31B Mg sheet. *Int. J. Plast.* **2007**, *23*, 44–86.

14. Barnett, M.R. Twinning and the ductility of magnesium alloys PartI: "Tension" twins. *Mater. Sci. Eng. A* **2007**, *464*, 1–7.

15. Kleiner, S.; Uggowitzer, P.J. Mechanical anisotropy of extruded Mg–6% Al–1% Zn alloy. *Mater. Sci. Eng. A* **2004**, *379*, 258–263.

16. Yi, S.B.; Davies, C.H.J.; Brokmeier, H.G.; Bolmaro, R.E.; Kainer, K.U.; Homeyer, J. Deformation and texture evolution in AZ31 magnesium alloy during uniaxial loading. *Acta Mater.* **2006**, *54*, 549–562.

17. El Kadiri, H.; Baird, J.C.; Kapil, J.; Oppedal, A.L.; Cherkaoui, M.; Vogel, S.C. Flow asymmetry and nucleation stresses of twinning and non-basal slip in magnesium. *Int. J. Plast.* **2013**, *44*, 111–120.

18. Oppedal, A.; Kadiri, H.E.; Tomé, C.; Vogel, S.C.; Horstemeyer, M. Anisotropy in hexagonal close-packed structures: Improvements to crystal plasticity approaches applied to magnesium alloy.*Philos. Mag.* **2013**, *93*, 4311–4330.

19. Bian, M.Z.; Shin, K.S. {10–12} twinning behavior in magnesium single crystal. *Met. Mater. Int.***2013**, *19*, 999–1004.

20. Al-Samman, T.; Molodov, K.D.; Molodov, D.A.; Gottstein, G.; Suwas, S. Softening and dynamic recrystallization in magnesium single crystals during c-axis compression. *Acta Mater.* **2012**, *60*, 537–545.

21. Molodov, K.D.; Al-Samman, T.; Molodov, D.A.; Gottstein, G. On the role of anomalous twinning in the plasticity of magnesium. *Acta Mater.* **2016**, *103*, 711–723.

22. Shin, J.H.; Kim, S.H.; Ha, T.K.; Oh, K.H.; Choi, I.S.; Han, H.N. Nanoindentation study for deformation twinning of magnesium single crystal. *Scr. Mater.* **2013**, *68*, 483–486.

23. Byer, C.M.; Li, B.; Cao, B.; Ramesh, K.T. Microcompression of single-crystal magnesium. *Scr. Mater.* **2010**, *62*, 536–539.

24. Molodov, K.D.; Al-Samman, T.; Molodov, D.A. On the diversity of the plastic response of magnesium in plane strain compression. *Mater. Sci. Eng. A* **2016**, *651*, 63–68.

25. Prasad, K.E.; Ramesh, K.T. In-situ observations and quantification of twin boundary mobility in polycrystalline magnesium. *Mater. Sci. Eng. A* **2014**, *617*, 121–126.

26. Kostopoulos, V.; Loutas, T.H.; Kontsos, A.; Sotiriadis, G.; Pappas, Y.Z. On the identification of the failure mechanisms in oxide/oxide composites using acoustic emission. *NDT E Int.* **2003**, *36*, 571–580.

27. Pappas, Y.Z.; Kontsos, A.; Loutas, T.H.; Kostopoulos, V. On the characterization of continuous fibres fracture by quantifying acoustic emission and acousto-ultrasonics waveforms. *NDT E Int.***2004**, *37*, 389–401.

28. Kontsos, A.; Loutas, T.; Kostopoulos, V.; Hazeli, K.; Anasori, B.; Barsoum, M.W. Nanocrystalline Mg-MAX composites: Mechanical behavior characterization via acoustic emission monitoring.*Acta Mater.* **2011**, *59*, 5716–5727.

29. Miguel, M.-C.; Vespignani, A.; Zapperi, S.; Weiss, J.; Grasso, J.-R. Intermittent dislocation flow in viscoplastic deformation. *Nature* **2001**, *410*, 667–671.

30. Vinogradov, A.; Hashimoto, S.; Miura, S. Effect of grain boundary on acoustic emission during plastic deformation of copper-aluminum bicrystals. *Acta Mater.* **1996**, *44*, 2883–2890.

31. Koslowski, M.; LeSar, R.; Thomson, R. Avalanches and scaling in plastic deformation. *Phys. Rev. Lett.* **2004**, *93*, 125502.

32. Ioka, I.; Yoda, S. Acoustic emission for polycrystalline graphite under compressive loading.*Carbon* **1990**, *28*, 489–495.

33. Chung, J.-B.; Kannatey-Asibu, E. Acoustic emission from plastic deformation of a pure single crystal. *J. Appl. Phys.* **1992**, *72*, 1812–1820.

34. Lamark, T.T.; Chmelík, F.; Estrin, Y.; Lukáč, P. Cyclic deformation of a magnesium alloy investigated by the acoustic emission technique. *J. Alloys Compd.* **2004**, *378*, 202–206.

35. Richeton, T.; Dobron, P.; Chmelik, F.; Weiss, J.; Louchet, F. On the critical character of plasticity in metallic single crystals. *Mater. Sci. Eng. A* **2006**, *424*, 190–195.

36. Máthis, K.; Čapek, J.; Zdražilová, Z.; Trojanová, Z. Investigation of tension–compression asymmetry of magnesium by use of the acoustic emission technique. *Mater. Sci. Eng. A* **2011**, *528*, 5904–5907.

37. Moorthy, V.; Jayakumar, T.; Raj, B. Acoustic Emission Teqcnique For Detecting Micro and Micro yeilding in solution annealed AISI Type 316 austenetic stainless steel. *Int. J. Pres. Ves. Piping* **1995**,*64*, 161–168.

38. Van Bohemen, S.M.C.; Sietsma, J.; Hermans, M.J.M.; Richardson, I.M. Kinetics of the martensitic transformation in low-alloy steel studied by means of acoustic emission. *Acta Mater.* **2003**, *51*, 4183–4196.

39. Vinogradov, A.; Merson, D.L.; Patlan, V.; Hashimoto, S. Effect of solid solution hardening and stacking fault energy on plastic flow and acoustic emission in Cu-Ge alloys. *Mater. Sci. Eng. A***2003**, *341*, 57–73.

40. Lu, Y.; Charghouri, M.; Taheri, F. Effect of texture on acoustic emission produced by slip and twinning in AZ31B magnesium alloy-part II: Clustering and neural network analysis. *Nondestruct. Test. Eval.* **2008**, *23*, 211–228.

41. Bohlen, J.; Chmelík, F.; Dobroň, P.; Letzig, D.; Lukáč, P.; Kainer, K.U. Acoustic emission during tensile testing of magnesium AZ alloys. *J. Alloys Compd.* **2004**, *378*, 214–219.

42. Meza-García, E.; Dobroň, P.; Bohlen, J.; Letzig, D.; Chmelík, F.; Lukáč, P.; Kainer, K.U. Deformation mechanisms in an AZ31 cast magnesium alloy as investigated by the acoustic emission technique. *Mater. Sci. Eng. A* **2007**, *462*, 297–301.

43. Muránsky, O.; Barnett, M.R.; Carr, D.G.; Vogel, S.C.; Oliver, E.C. Investigation of deformation twinning in a fine-grained and coarse-grained ZM20 Mg alloy: Combined in situ neutron diffraction and acoustic emission. *Acta Mater.* **2010**, *58*, 1503–1517.

44. Bohlen, J.; Dobron, P.; Meza, G.E.; Chmelík, F.; Lukáč, P.; Letzig, D.; Kainer, K.U. The Effect of Grain Size on the Deformation Behaviour of Magnesium Alloys Investigated by the Acoustic Emission Technique. *Adv. Eng. Mater.* **2006**, *8*, 422–427.

45. Bohlen, J.; Dobroň, P.; Swiostek, J.; Letzig, D.; Chmelík, F.; Lukáč, P.; Kainer, K.U. On the influence of the grain size and solute content on the AE response of magnesium alloys tested in tension and compression. *Mater. Sci. Eng. A* **2007**, *462*, 302–306.

46. Vinogradov, A.; Orlov, D.; Danyuk, A.; Estrin, Y. Effect of grain size on the mechanisms of plastic deformation in wrought Mg–Zn–Zr alloy revealed by acoustic emission measurements. *Acta Mater.* **2013**, *61*, 2044–2056.

47. Drozdenko, D.; Bohlen, J.; Chmelík, F.; Lukáč, P.; Dobroň, P. Acoustic emission study on the activity of slip and twin mechanisms during compression testing of magnesium single crystals.*Mater. Sci. Eng. A* **2016**, *650*, 20–27.

48. Čapek, J.; Máthis, K.; Clausen, B.; Stráská, J.; Beran, P.; Lukáš, P. Study of the loading mode dependence of the twinning in random textured cast magnesium by acoustic emission and neutron diffraction methods. *Mater. Sci. Eng. A* **2014**, *602*, 25–32.

49. Drozdenko, D.; Bohlen, J.; Yi, S.; Minárik, P.; Chmelík, F.; Dobroň, P. Investigating a twinning–detwinning process in wrought Mg alloys by the acoustic emission technique. *Acta Mater.* **2016**,*110*, 103–113.

50. Hazeli, K.; Askari, H.; Cuadra, J.; Streller, F.; Carpick, R.W.; Zbib, H.M.; Kontsos, A. Microstructure-sensitive investigation of magnesium alloy fatigue. *Int. J. Plast.* **2015**, *68*, 55–76.

51. Kawai, S.; Yoshida, Y.; Asano, H.; Miura, S.; Kobayashi, M.; Shibano, J.-I. Acoustic emission response of magnesium alloy during cyclic and creep tests. *Mater. Sci. Eng. A* **2016**, *668*, 120–124.

52. Chang, H. Identification of damage mode in AZ31 magnesium alloy under tension using acoustic emission. *Trans. Nonferr. Metal. Soc.* **2015**, *25*, 1840–1846.

53. Vinogradov, A.; Orlov, D.; Danyuk, A.; Estrin, Y. Deformation mechanisms underlying tension–compression asymmetry in magnesium alloy ZK60 revealed by acoustic emission monitoring.*Mater. Sci. Eng. A* **2015**, *621*, 243–251.

54. Vinogradov, A.; Lazarev, A.; Linderov, M.; Weidner, A.; Biermann, H. Kinetics of deformation processes in high-alloyed cast transformation-induced plasticity/twinning-induced plasticity steels determined by acoustic emission and scanning electron microscopy: Influence of austenite stability on deformation mechanisms. *Acta Mater.* **2013**, *61*, 2434–2449.

55. Linderov, M.; Segel, C.; Weidner, A.; Biermann, H.; Vinogradov, A. Deformation mechanisms in austenitic TRIP/TWIP steels at room and elevated temperature

investigated by acoustic emission and scanning electron microscopy. *Mater. Sci. Eng. A* **2014**, *597*, 183–193.

56. Hazeli, K.; Cuadra, J.; Streller, F.; Barr, C.M.; Taheri, M.L.; Carpick, R.W.; Kontsos, A. Three-dimensional effects of twinning in magnesium alloys. *Scr. Mater.* **2015**, *100*, 9–12.

57. Hazeli, K.; Cuadra, J.; Vanniamparambil, P.A.; Kontsos, A. In situ identification of twin-related bands near yielding in a magnesium alloy. *Scr. Mater.* **2013**, *68*, 83–86.

58. Hazeli, K.; Sadeghi, A.; Pekguleryuz, M.; Kontsos, A. The effect of Strontium in Plasticity of Magnesium alloys. *Mater. Sci. Eng. A* **2013**, *578*, 383–393.

59. Wisner, B.; Cabal, M.; Vanniamparambil, P.A.; Hochhalter, J.; Leser, W.P.; Kontsos, A. In Situ Microscopic Investigation to Validate Acoustic Emission Monitoring. *Exp. Mech.* **2015**, *55*, 1705–1715. Vanniamparambil, P.A.; Guclu, U.; Kontsos, A. Identification of Crack Initiation in Aluminum Alloys using Acoustic Emission. *Exp. Mech.* **2015**, *55*, 837–850.

60. Yu, Q.; Jiang, Y.; Wang, J. Cyclic deformation and fatigue damage in single-crystal magnesium under fully reversed strain-controlled tension–compression in the [10–10] direction. *Scr. Mater.***2015**, *96*, 41–44.

61. Baxevanakis, K.P.; Cabal, M.; Kontsos, A. The Role of Shear Bands in Mechanical Behavior of Magnesium Alloys. *Int. J. Soids Struct.* **2016**. under review.

62. Aydıner, C.C.; Telemez, M.A. Multiscale deformation heterogeneity in twinning magnesium investigated with in situ image correlation. *Int. J. Plast.* **2014**, *56*, 203–218.

63. Barnett, M.R.; Nave, M.D.; Ghaderi, A. Yield point elongation due to twinning in a magnesium alloy. *Acta Mater.* **2012**, *60*, 1433–1443.

64. Li, Y.; Enoki, M. Twinning behavior of pure magnesium quantitatively investigated by acoustic emission. *Mater. Sci. Eng. A* **2012**, *536*, 8–13.

65. Máthis, K.; Chmelík, F.; Janeček, M.; Hadzima, B.; Trojanová, Z.; Lukáč, P. Investigating deformation processes in AM60 magnesium alloy using the acoustic emission technique. *Acta Mater.* **2006**, *54*, 5361–5366.

66. Tou, J.T. A Dynamic Optimal Cluster-seeking Technique. *Int. J. Comput. Int. Sci.* **1979**, *8*, 541–547.

67. Cabal, M. Microstructurally-driven investigation of strain localization in wrought magnesium alloys. In *Mechanical Engineering & Mechanics*; Drexel University: Philadelphia, PA, USA, 2015.

© 2016 by the authors; licensee MDPI, Basel, Switzerland. This article is an open access article distributed under the terms and conditions of the Creative Commons Attribution (CC-BY) license (http://creativecommons.org/licenses/by/4.0/).

6

CHAPTER

Experimental Ultrasound Transmission through Fluid-Solid and Air-Solid Phononic Plates

Vicente Gómez-Lozano, Constanza Rubio *, Pilar Candelas, Antonio Uris and Francisco Belmar

Centro de Tecnologías Físicas, Universitat Politécnica de Valencia, Avd. Los Naranjos, s/n., Valencia 46022, Spain

Abstract

Underwater ultrasonic transmissions for fluid-solid and air-solid phononic brass plates are reported in this work. Although the structure is roughly the same, experimental results show very different behaviour between fluid-solid and air-solid phononic plates, due to most of the properties of the fluid-solid perforated plates rely on Fabry-Perot resonances, Wood anomalies and Lamb modes. In air-solid phononic plates Fabry-Perot resonance is highly attenuated due to impedances difference between air and water, and therefore some transmission modes are now distinguishable due to surface modes coupling.

Keywords: subwalength hole array; perforated plate; phononic plate; Lamb waves

PACS:

43.20.+g; 43.35.+d; 42.79.Dj

1. Introduction

At the end of the 19th century Lord Rayleigh [1] studied the reflection coefficient of a one-dimensional grating. This was the first study of sound interaction with periodic structures. Since then there have been major studies of sound wave propagation through periodic structures. During the last decades, much attention has been paid to phononic crystal. The beginning of the research in the field of phononic crystals started nearly two decades ago considering mainly bulk wave propagation [2,3]. Concepts such as band-gaps, waveguiding, and negative refraction among others rapidly extended to plate guided waves (Lamb waves). Theoretical [4,5,6] and experimental [7,8,9] studies dealt firstly with band gap and waveguiding phenomena for air-solid and solid-solid phononic plates. Most of the work done so far for phononic plates restricts the sound wave propagation to the solid and the effect of the fluid (which could be treated as vacuum) is then neglected. Recently, phononic crystal strips are applied to enhance the quality factor of aluminum nitride (AlN) electroacoustic resonators utilizing the lowest symmetric (S_0) Lamb wave mode thanks to strong reflection in acoustic band gaps [10].

On the other hand, a periodically perforated plate structure has been brought into focus while researching on the acoustic counterpart of Extraordinary Optical Transmission [11,12]. Resonant full transmission [13,14], Wood anomalies [15,16], and non-leaky modes [17,18] have been reported neglecting the role of the solid and constraining the wave propagation to the fluid. These two complementary points of view can be combined via fluid-solid coupling. Although more difficult to deal with, this unified perspective enables the observation of Extraordinary Sound Screening [19], complex interactions between Lamb-like modes and hole resonances [20,21], and full transmission through non perforated (corrugated) plates [22]. It is clear that the fluid-solid coupling plays a crucial role in the acoustical properties of phononic plates, portraying a wide range of different behaviours.

In this letter we provide experimental results for fluid loaded fluid-solid and air-solid phononic plates. Angle resolved transmission measurements of underwater ultrasound reveal key aspects of the different behaviour of phononic plates having fluid or air inclusions.

2. Materials and Methods

Underwater ultrasound has been chosen to measure the acoustic transmission through perforated plates. The geometrical parameters to describe perforated plates are the diameter of the hole d, the array periodicity a and the plate thickness h. The experimental setup is based on the well-known ultrasonic immersion transmission technique. This technique makes use of a couple of transmitter/receiver ultrasonic transducers. Two different couples of transducers were used: one couple with a center frequency of 250 kHz and a frequency range between 155–350 kHz and the other couple with center frequency of 500 kHz and a frequency range between 350 and 650 kHz. Each transducer was located at a distance larger than that of its nearfield distance (43 mm) from the plate and aligned with respect to the plate. A pulse is launched by the emitter piston transducer through the inspected plate. Then, the signal is detected by the receiving piston transducer and acquired by the pulser/receiver, post amplified and digitized by a digital PC oscilloscope (Picoscope model 3224, Cambridge, UK). Time domain data is finally analyzed after averaging 100 different measures and deleting unwanted reflections by means of a time window. The transmission spectrum is then calculated from the power spectrum of the signal normalized with the reference signal power spectrum measured without the sample plate. Typically, angle dependent measurements were done in angle steps of 1° and comprising from 0° to 60°.

The measurements were made using brass plates with 350 mm in width and 450 mm in length (ρ = 7890 kg/m^3, c_l = 5670 m/s, c_t =

3230 m/s) and 2 mm thickness, immersed in water (ρ = 1000 kg/m^3, c_l = 1480 m/s).

A brass plate drilled with periodically square distribution of circular holes having diameter, *d*, of 1 mm and a unit cell period, *a*, of 3 mm, was used (Figure 1). Starting from the perforated plate the fabrication of an air-solid phononic plate was carried out by sticking, carefully, a film (20 μm thick) around the plate, taking care that did not appear undesired air bubbles when the film was adhered.

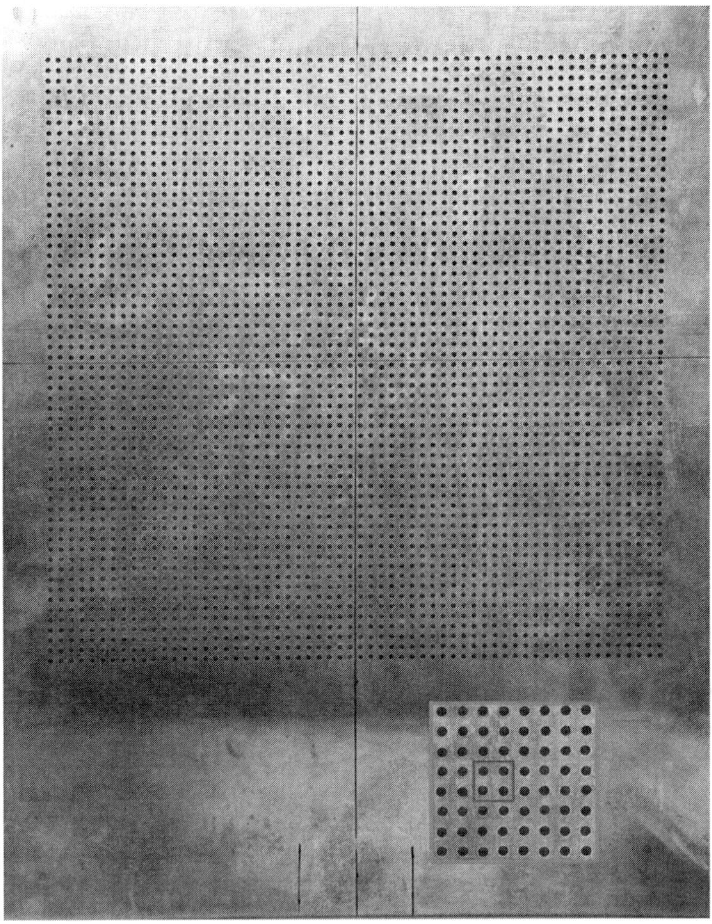

Figure 1. Structure and elementary cell for a brass plate perforated with periodically square distribution of circular holes.

3. Results and Discussion

When an incident sound pressure wave impinges on the plate, gives rise a reflected and a transmitted sound pressure wave. In the solid, high transmission values can be observed revealing a complex dispersion. For this case, three kinds of modes can be distinguished: Scholte-Stoneley mode, symmetric leaky Lamb modes, and antisymmetric leaky Lamb modes. The Scholte-Stoneley mode propagates across the fluid-solid interface with a phase speed slightly slower than that of the water. At low frequencies, this mode is mixed with the A_0 mode and slowly converges to the sound line as ω increases. Leaky Lamb modes are guided waves produced due to the multiple reflections of longitudinal and in-plane transverse modes at both plate-fluid interfaces. Figure 2 shows the different shapes of the symmetric and antisymmetric modes in a 2 mm thickness brass plate. Although there was not periodicity due to it was not still drilled, in order to compare with subsecuent results, these are plotted in terms of the normalized frequencies $k_0 \cdot a/\pi$ and $k_{//} \cdot a/\pi$, where a is the array periodicity of the perforated plate, where k_0 is the incident wavenumber and $k_{//}$ is k_0 projection on the plate.

However, when the plate is periodically perforated, one has to consider the elastic movement of the plate coupled with the surrounding fluid not only at the plate free surface but also inside the apertures. In order to evaluate the physical phenomena, without taking into account the elastic movement of the plate, the transmission sound power can be calculated by solving the sound wave equation within a hard-solid model [23], that is, assuming that the pressure field does not penetrate into the material plate because the water/plate impedance is considered infinite. Figure 3 shows the calculated transmitted sound power coefficient, τ, for the periodically square distribution of circular holes considered and it is observed the Fabry-Perot full transmission at $k_0 \cdot a/\pi = 1.2$ and the Wood anomaly minima at $k_0 \cdot a/\pi = 2$ that is evident when the incidence angle is varied. Due to the elastic movement of the plate had not been taken into account, Lamb and Scholte-Stoneley waves are not observed.

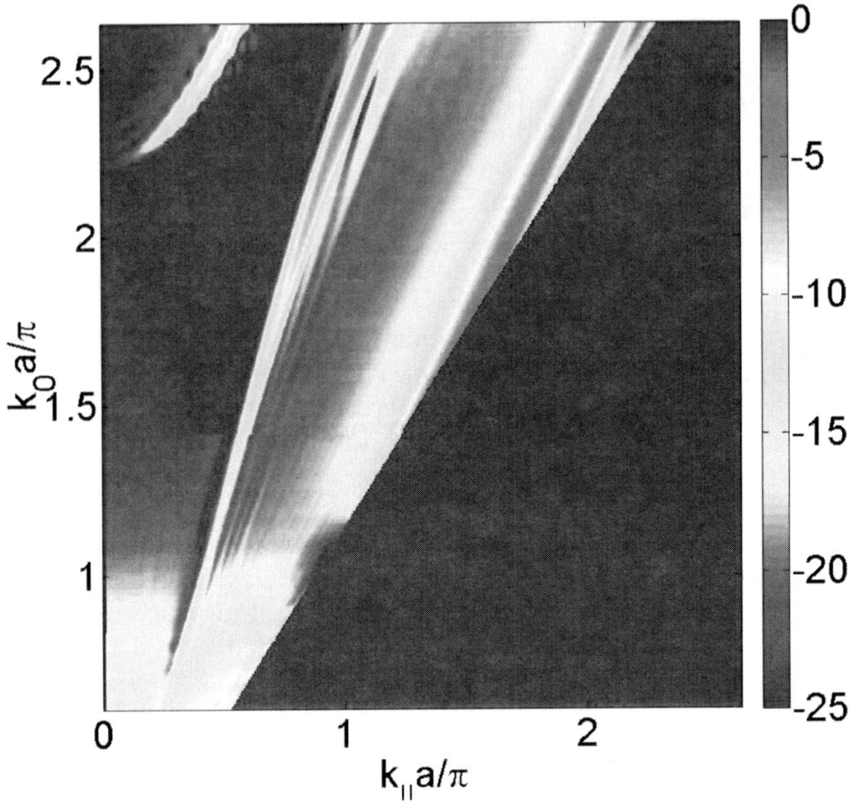

Figure 2. Measured transmitted sound power coefficient, τ, as a function of the parallel wavevector $k_{//}\cdot a/\pi$ in the ΓX direction and the normalized frequency $k_0\cdot a/\pi$, where *a* the array periodicity. The transmitted sound is represented in dB scale.

Figure 4 depicts the experimental transmitted sound power τ, as a function of the parallel wavevector$k_{//}\cdot a/\pi$ in the ΓX direction and the normalized frequency $k_0\cdot a/2\pi$, of the brass plate perforated of thickness h = 2 mm with holes of diameter d = 1 mm arranged periodically in a square lattice of period a= 3 mm and immersed in water. Complex interaction between minima and maxima is present in the spectra and it makes clear that the symmetry of the array results in a high angular dependence of the spectra. For $k_{//}\cdot a/\pi$ near zero (normal incidence) the Fabry-Perot full transmission at $k_0 \cdot a/\pi$ = 1.2 and Wood anomaly can be clearly seen both thickness at $k_0 \cdot a/\pi \approx 2$. However, the modes appearing from the bottom of the figures are not predicted by the Wood anomaly and are related to leaky surface modes. These surface modes are leaky Lamb modes observed in the homogeneous plate (see Figure 2) arising from the plate vibration and the solid-fluid coupling.

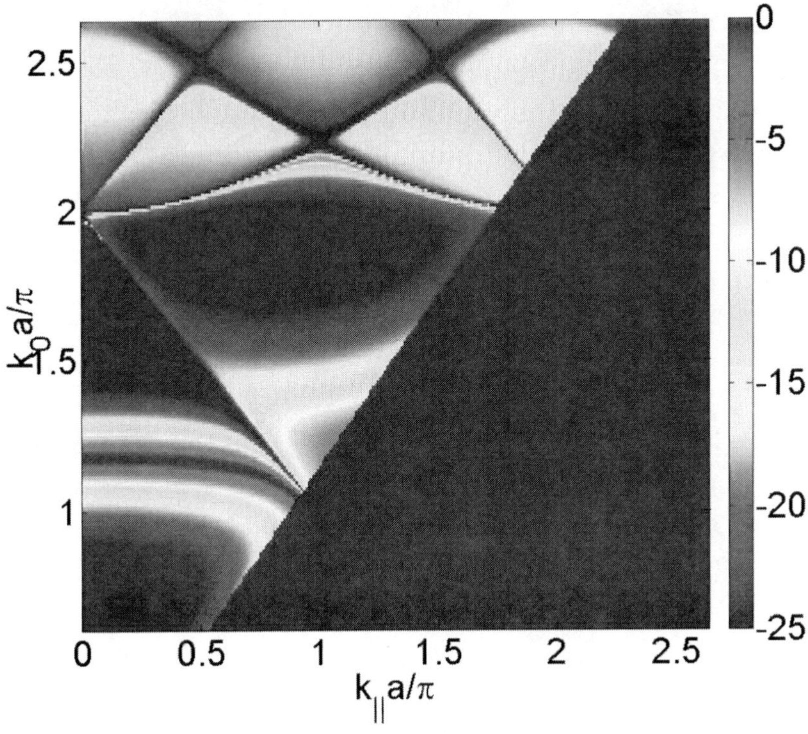

Figure 3. Calculated transmitted sound power coefficient, τ, as a function of the parallel wavevector $k_{//} \cdot a/\pi$ in the Γ*X* direction and the normalized frequency $k_0 \cdot a/\pi$, for the square distribution of circular holes having diameter, *d*, of 1 mm and a unit cell period, *a*, of 3 mm. The transmitted sound is represented in dB scale.

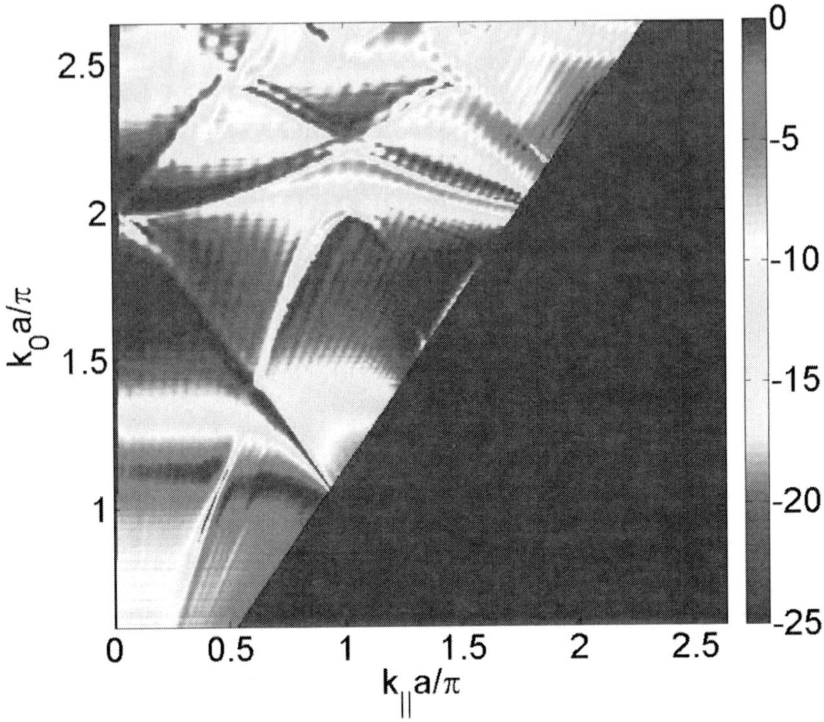

Figure 4. Measured transmitted sound power coefficient, τ, as a function of the parallel wavevector $k_{//}\cdot a/\pi$ in the Γ*X* direction and the normalized frequency $k_0\cdot a/\pi$, for the square distribution of circular holes having diameter, *d*, of 1 mm and a unit cell period, *a*, of 3 mm. The transmitted sound is represented in dB scale. The transmitted sound is represented in dB scale.

The experimental dispersion of the same brass plate but having air filled holes is showed in Figure 5. As the wave interaction at an air-solid interface is different from that of a fluid-solid interface, the transmission properties of the phononic plate differ from the previous case. As the holes have been filled with air, the Fabry-Perot resonance was highly attenuated due to the impedance difference between water and air. Zero order Lamb modes, Scholte-Stoneley mode near grazing incidence and Wood anomalies are clearly distinguishable but there are new transmission modes that have not been detected in the case of the water filled holes.

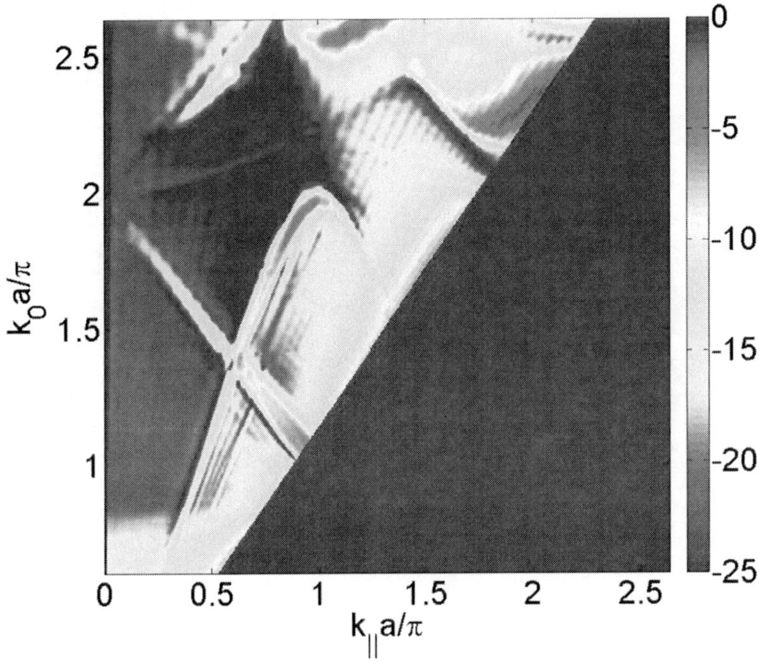

Figure 5. Measured transmitted sound power coefficient, τ, as a function of the parallel wavevector $k_{//}\cdot a/\pi$ in the Γ*X* direction and the normalized frequency $k_0\cdot a/\pi$, for the square distribution of circular holes having diameter, *d*, of 1 mm and a unit cell period, *a*, of 3 mm and the holes filled with air (air-solid phononic plate). The transmitted sound is represented in dB scale.

The nature of these transmission modes can be demonstrated by calculating the dispersion diagram for the brass phononic plate. For this purpose we calculate the eigenfrequencies of the cell type with periodic boundary conditions of Floquet type with a variable vector $\vec{k}$ =(k,0,0) between 0 and $\pi/2a$.Figure 6 shows the dispersion diagram.

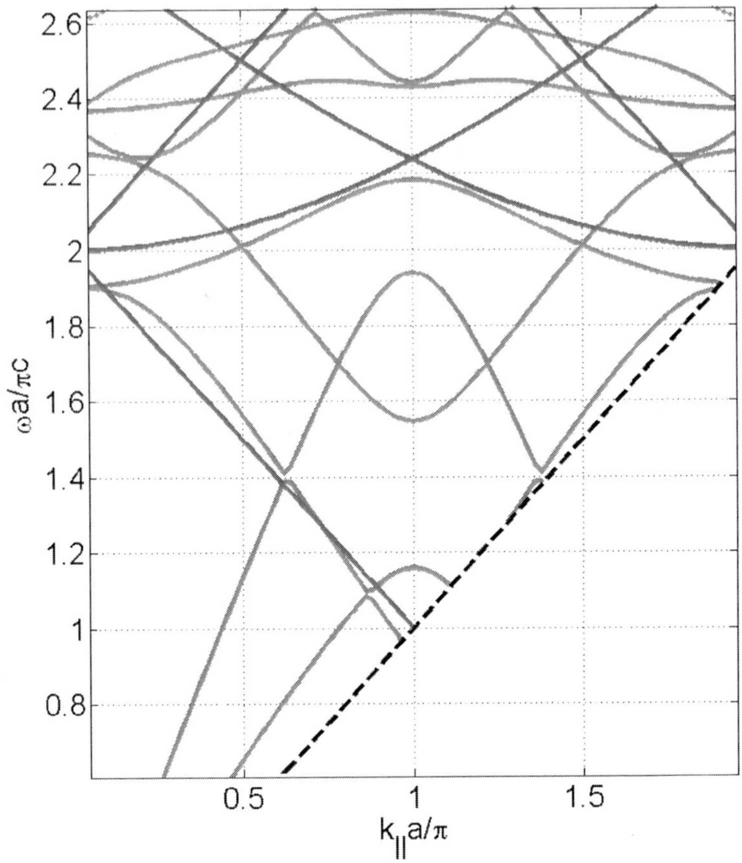

Figure 6. Calculated dispersion diagram (red lines) and Wood minima (blue lines), as a function of the parallel wavevector $k_{//}\cdot a/\pi$ in the ΓX direction and the normalized frequency $\omega a/\pi c$, for the square distribution of circular holes having diameter, *d*, of 1 mm and a unit cell period, *a*, of 3 mm.

By superimposing the dispersion diagram with the graphics of the transmitted sound power coefficient as a function of the parallel wavevector for the perforated plates with water or air-filled holes (see Figure 7), it is observed that while some lines of the dispersion diagram clearly indicate a significant increase in transmission through plate, others do not exhibit.

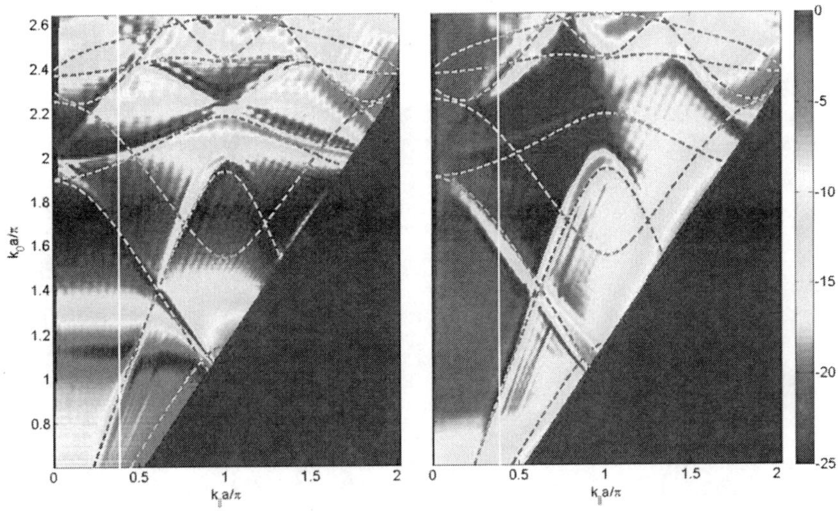

(a) (b)

Figure 7. Dispersion diagram superimposed to measured transmitted sound power coefficient, τ, as a function of the parallel wavevector $k_{//}\cdot a/\pi$ in the ΓX direction and the normalized frequency$k_0\cdot a/\pi$, for the square distribution of circular holes having diameter,d, of 1 mm and a unit cell period, a, of 3 mm and the holes filled with (**a**) water and (**b**) air.

To analyze this fact, we calculate the displacement of the unit cell in the z direction for a fixed value of the parallel wavevector $k_{//}\cdot a/\pi$ for the corresponding eigenfrequency. Figure 8 shows some eigenmodes for $k_{//}\cdot a/\pi = 0.4$. It is noted that some are antisymmetric modes (A_1, A_2, A_3 and A_4) with respect to the forward direction of the plane wave. Clearly these eigenmodes produce poor fluid-solid coupling and lead to non-radiative bands, while symmetric eigenmodes (S_1, S_2, S_3 and S_4) produce defined transmission lines.

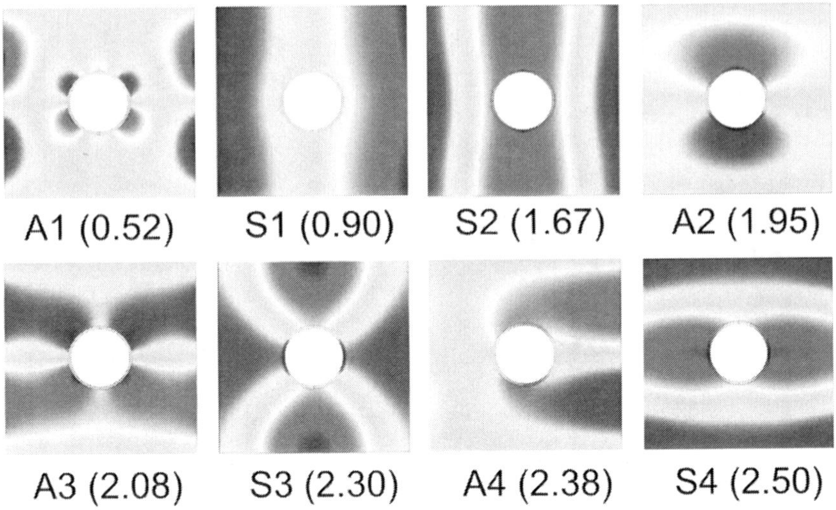

Figure 8. z-displacement of the brass unit cell for a fixed value of$k_{//}\cdot a/\pi$ = 0.4 The eigenmodes are labeled as S (symmetric) or A(antisymmetric) according to they behaviour to the horizontal plane wave incoming from the left side of the plots. The eigenmodes normalized frequency $k_0\cdot a/\pi$ values are despited in parenthesis. The displacement is represented in linear scale.

4. Conclusions

The results presented in this letter give a wide description of the phenomena involved in phononic plates as they also includes the coupling with the surrounding fluid. The fluid-solid phononic plates behavior is mainly governed by holes resonances and coherent diffraction effects which interact with Lamb-like modes. However, in air-solid phononic plates, zero order Lamb modes, Scholte-Stoneley mode near grazing incidence, Wood anomalies and new transmission modes are clearly distinguishable while the Fabry-Perot resonance was highly attenuated due to the impedance difference between water and air. It has been reported here that these differences are due to antisymmetric eigenmodes produce poor fluid-solid coupling and lead to non-radiative modes, while symmetric eigenmodes has high fluid-solid coupling and lead radiative modes.

Acknowledgments

This work has been supported by Spanish MINECO (TEC2015-70939-R) and Generalitat Valenciana (AICO/2015/119).

Author Contributions

Constanza Rubio and Antonio Uris has coordinated the theoretical and experimental developments, participating in the establishment of the theory principles used in this work, as well as in the drafting of the manuscript. Vicente Gomez-Lozano has developed part of the theory used and has designed some characterization experiments and has developed some computing task, as well as has participated in the drafting of the manuscript. Pilar Candelas, Francisco Belmar have designed and carried out the characterization of the experimental setup and in computing algorithms for the treatment of data, also have participated in the analysis of the state of art and in the design of some experiments, as well as in the drafting of the manuscript.

Conflicts of Interest

The authors declare no conflict of interest. The founding sponsors had no role in the design of the study; in the collection, analyses, or interpretation of data; in the writing of the manuscript, and in the decision to publish the results.

References

1. Rayleigh, L. *The Theory of Sound*, 2nd ed.; Courier Dover Publications: New York, NY, USA, 1945; Volume 2, p. 89.

2. Miyashita, T. Sonic crystals and sonic waveguides. *Meas. Sci. Technol.* **2005**, *16*, R47–R63. Sigalas, M.; Kushwaha, M.S.; Economou, E.N.; Kafesaki, M.; Psarobas, I.E.; Steurer, W. Classical vibrational modes in photonic lattices: Theory and experiment. *Z. Kristallogr.* **2005**, *220*, 765–809.

3. Chen, J.-J.; Qin, B.; Cheng, J.C. Complete band gaps for Lamb waves in cubic thin plates with periodically placed inclusions. *Chin. Phys. Lett.* **2005**, *22*, 1706–1708.

4. Khelif, A.; Aoubiza, B.; Mohammadi, S.; Adibi, A.; Laude, V. Complete band gaps in two-dimensional phononic crystal slabs. *Phys. Rev. E* **2006**, *74*, 045510.

5. Vasseur, J.O.; Deymier, P.A.; Djafari-Rouhani, B.; Pennec, Y.; Hladky-Hennion, A.C. Absolute forbidden bands and waveguiding in two-dimensional phononic crystal plates. *Phys. Rev. B* **2008**,*77*, 085415.

6. Zhang, X.; Jackson, T.; Lafond, E.; Deymier, P.; Vasseur, J. Evidence of surface acoustic wave band gaps in the phononic crystals created on thin plates. *Appl. Phys. Lett.* **2006**, *88*, 041911.

7. Brunet, T.; Vasseur, J.; Bonello, B.; Djafari-Rouhani, B.; Hladky-Hennion, A.C. Lamb waves in phononic crystal slabs with square or rectangular symmetries. *J. Appl. Phys.* **2008**, *104*, 043506.

8. Mohammadi, S.; Eftekhar, A.A.; Khelif, A.; Hunt, W.D.; Adibi, A. Evidence of large high frequency complete phononic band gaps in silicon phononic crystal plates. *Appl. Phys. Lett.* **2008**,*92*, 221905.

9. Lin, C.-M.; Hsu, J.-C.; Senesky, D.G.; Pisano, A.P. Anchor loss reduction in AlN Lamb wave resonators using phononic crystal strip tethers. *Proc. IEEE Int. Freq. Control Symp.* **2014**.

10. Ebbesen, T.W.; Lezec, H.J.; Ghaemi, H.F.; Thio, T.; Wolf, P.A. Extraordinary optical transmission through subwavelength hole arrays. *Nature* **1998**, *391*, 667–669.

11. Garcia-Vidal, F.J.; Martin-Moreno, L.; Ebbesen, T.W.; Kuipers, L. Light passing through subwavelength apertures. *Rev. Mod. Phys.* **2010**, *82*, 729–787.

12. Zhou, L.; Kriegmann, G.A. Complete transmission through a periodically perforated rigid slab. *J. Acoust. Soc. Am.* **2007**, *121*, 3288–3299.

13. Lu, M.; Liu, X.; Feng, L.; Li, J.; Huang, C.; Chen, Y.; Zhu, Y.; Zhu, S.; Ming, N. Extraordinary acoustic transmission through a 1D grating with very narrow apertures. *Phys. Rev. Lett.* **2007**, *99*, 174301.

14. Hou, B.; Mei, J.; Ke, M.; Wen, W.; Liu, Z.; Shi, J.; Sheng, P. Tuning Fabry-Perot resonances via diffraction evanescent waves. *Phys. Rev. B* **2007**, *76*, 054303.

15. Wood, R.W. On a remarkable case of uneven light in a diffraction grating. *Philos. Mag.* **1902**, *4*, 396–402.

16. Christensen, J.; Martin-Moreno, L.; Garcia-Vidal, F.J. Theory of resonant acoustic transmission subwavelength apertures. *Phys. Rev. Lett.* **2008**, *101*, 014301.

17. He, Z.; Jia, H.; Qiu, C.; Ye, Y.; Hao, R.; Ke, M.; Liu, Z. Nonleaky surface acoustic waves on a textured rigid surface. *Phys. Rev. B* **2011**, *83*, 132101.

18. Estrada, H.; Candelas, P.; Uris, A.; Belmar, F.; Garcia de Abajo, F.J.; Meseguer, F. Extraordinary sound screening in perforated plates. *Phys. Rev. Lett.* **2008**, *101*, 084302.

19. Estrada, H.; Candelas, P.; Uris, A.; Belmar, F.; Garcia de Abajo, F.J.; Meseguer, F. Influence of lattice symmetry on ultrasound transmission through plates with subwavelength aperture arrays.*Appl. Phys. Lett.* **2009**, *95*, 051906.

20. Estrada, H.; Garcia de Abajo, F.J.; Candelas, P.; Uris, A.; Belmar, F.; Meseguer, F. Angle-dependent ultrasonic transmission through plates with subwavelength hole arrays. *Phys. Rev. Lett.* **2009**, *102*, 144301.

21. He, Z.; Jia, H.; Qiu, C.; Peng, S.; Mei, X.; Cai, F.; Peng, P.; Ke, M.; Liu, Z. Acoustic transmission enhancement through a

periodically structured stiff plate without any opening. *Phys. Rev. Lett.* **2010**, *105*, 074301.

22. Estrada, H.; Candelas, P.; Uris, A.; Belmar, F.; Meseguer, F.; Garcia de Abajo, F.J. Sound transmission through perforated plates with subwavelength hole arrays: A rigid-solid model. *Wave Motion* **2011**, *48*, 235–242.

© 2016 by the authors; licensee MDPI, Basel, Switzerland. This article is an open access article distributed under the terms and conditions of the Creative Commons Attribution (CC-BY) license (http://creativecommons.org/licenses/by/4.0/).

7

CHAPTER

First Evidence of Surface SH-Wave Propagation in Cubic Piezomagnetics

Aleksey Zakharenko

International Institute of Zakharenko Waves (IIZWs), Krasnoyarsk, Russia.

ABSTRACT

This theoretical work provides with results of characteristics calculation of the ultrasonic surface Zakharenko waves (USZWs) existing in piezomagnetic cubic monocrystals of class m3m that can be readily used for non-destructive testing. The piezomagnetic waves propagate in direction [101] corresponding to relatively easy magnetization for the following piezomagnetics: Galfenol, Terfenol-D, and $CoFe_2O_4$ with cubic structures. The phase velocities of the USZW-waves and the coefficient of magnetomechanical coupling (CMMC) K^2 were calculated for the crystals. It was found that the coefficient K^2 for piezomagnetics with $K_m^2 > 1/3$ and $K_m^2 >> 1/3$ is about 8% to 9%, where $K^2 = 2\ (V_{USZW,o} - V_{USZW,c})/V_{USZW,o}$ and $K_m^2 = h^2/(C\mu)$. Knowledge of piezomagnetic properties of cubic crystals makes possible the use of them in new products utilizing the phenomenon called the magnetoelectric effect. Also, this study is useful for possible application of cubic piezomagnetics in composite structures consisting of piezoelectric and (or) piezomagnetic materials and in the microwave technology. This broadens choice of possible piezomagnetic materials for utilization in various technical devices.

Keywords: piezomagnetic cubic monocrystals, Galfenol and Terfenol-D, ultrasonic surface Zakharenko waves

1. Introduction

Ferroic materials can display a spontaneous magnetization (ferromagnetic), polarization (ferroelectric) and strain (ferroelastic, shape memory alloy). Materials possessing two "ferro" properties simultaneously are called "biferroics". There are three kinds of biferroic materials: electroelastic, magnetoelastic, and magnetoelectric materials. The magnetoelectric materials (composites) can simultaneously possess ferromagnetic and ferroelectric properties. Composite materials or structures consisting of piezoelectric and piezomagnetic phases are able to facilitate the conversion of energy between electric and magnetic fields. Such phenomenon is called magnetoelectric (ME) effect. The ME effect of the composites was first reported by van Suchtelen [1]. The ME effect was then studied by van den Boomgaard [2], van Run et al. [3], and van den Boomgaard et al. [4] for $BaTiO_3$/ $CoFe_2O_4$ composites. Possible applications of ME materials include magnetic-electric energy converting components; solid state non-volatile memory, multi-state memory, which can find application in quantum computing area; and electrical/optical polarization components, which can find applications in communication, light computing, and solid state memories based on spintronics [5,6]. Also, this research topic can be found in a recent review paper [7] by Fiebig. Note that the ME effect was originally predicted by P. Curie [8] in 1894.

Surface acoustic wave (SAW) devices are widely used in numerous branches of science and technology and their investigation especially in the case of interconnected physical fields is an important and actively developing branch of research and applications. About 120 years ago the first type of SAW was described by Lord Rayleigh [9] in connection with the problem of earthquakes. Bleustein [10] and Gulyaev [11] theoretically predicted that a pure shear-horizontal (SH) surface wave can be guided by the free surface of a piezoelectric half-space. Later, Maerfeld and Tournois [12] investigated SH acoustic waves guided by the interface of two half-spaces, and Danicki [13] described the acoustic waves which can be guided by an embedded conducting plane in the electro-elastic materials of class 6 mm. It is thought that the simplest type of SH-SAW represents Love waves [14] in layered systems, consisting of an isotropic layer and an isotropic substrate. Recent developments in physics and technology made possible to construct new magneto-electro-elastic materials which demonstrate interconnection between magnetic, electricand elastic fields [15,16]. When the electric field was connected with the elastic one (piezoelectric materials) it brought up new and unexpected possibilities for science and technology. Connecting the magnetic field with the electric and elastic ones in magneto-electro-elastic materials suggests a range of new possibilities. Alshits et al. [17] conducted a qualitative investigation on the existence of surface waves in half-infinite anisotropic elastic media with piezoelectric, piezomagnetic, and ME effects. Soh and Liu [18] gave the existence conditions of interfacial SH-waves in a piezoelectric/piezomagnetic bi-material. Although much attention has been recently concentrated on magneto-electro-elastic materials and several dynamic problems have been solved by Hu and Li [19], Li [20], Chen, et al., [21], etc., the investigation of SAW propagation in transversely isotropic magneto-electro-elastic materials [22-25] of class 6 mm is currently an open actual subject. In addition to the transversely isotropic materials there are piezoelectric and piezomagnetic crystals with cubic symmetries, in which the surface BG-waves [10,11] cannot exist, according to [26] by Gulyaev and Hickernell. That is also true for the interfacial MTwaves [12]. However, new SH-SAWs called the ultrasonic surface Zakharenko waves (USZWs) were recently discovered in [27], which can propagate in cubic piezoelectrics. That is also true for cubic piezomagnetics. The purpose of this paper is to show the first evidence of the USZW existence in cubic piezomagnetics. It is possible to introduce piezomagnetic cubic crystals which are today highly-called for different technical devices.

The Cobalt Ferrite $CoFe_2O_4$ ($CoO \cdot Fe_2O_3$ or CFO) has the spinel ferrite structure with a stoichiometry denoted by AB_2O_4, where A is a divalent transition metal ion or a mixture of a trivalent and a monovalent metal ion. A spinel (see figure 1 and [28]) is a cubic structure in which O^{2-} ions form an fcc-lattice. It requires eight Formula equivalent AB_2O_4 to form a repeating unit cell which contains 32 O^{2-} ions, 64 tetragonal sites (A sites) and 32 octahedral sites (B sites). However, only 8 tetragonal and 16 octahedral sites are occupied so that the ratio of A-atoms/B-atoms is 1:2. CFO is known to have highly anisotropic magnetic properties and has been widely used in magnetic media and microwave device applications.

In 1971, the U.S. Navy developed an alloy of rare metals Tb and Dy with iron to create Terfenol-D ($Tb_xDy_{1-x}Fe_y$). The name "Terfenol-D" comes from a combination of TER for terbium, FE for iron, NOL for Naval Ordnance Laboratory, and D for dysprosium. Recently an iron/gallium alloy termed Galfenol has also been discovered at the Naval Surface Warfare Center (former NOL), where the magnetostrictive material Terfenol-D was discovered. It was found that magnetostriction in iron/gallium alloys peaks at a volume fraction of ~17% gallium. Unlike Terfenol-D representing a giant magnetostrictive material useable in practical operating conditions, FeGa-alloys are tough and not toxic, and can be machined and used without any special handling in devices where Terfenol-D may fracture. The cubic structure of Galfenol is shown in figure 2. Terfenol-D has been commercially available since the late 1980's and has since grown into an international industry. The Terfenol-D rods can be readily manufactured using the freestand-zone-melt process resulting in a cubic crystal structure [29] featuring a large magnetic anisotropy. Also, FeGa-alloys appear promising in filling the role as a mechanically robust material with substantial magnetostrictive capability [30]. Additional benefits of FeGa-alloys include substantially lower material costs than rare-earth based magnetostrictive alternatives. For example, the Terfenol-D ($Tb_{27}Dy_{73}Fe_{195}$) raw material cost is approximately \$ 0.50/g and current production techniques utilize carefully controlled crystal growth processes [31]. For comparison, the raw material cost for $Fe_{81}Ga_{19}$ is approximately \$ 0.08/g [31]. Indeed,

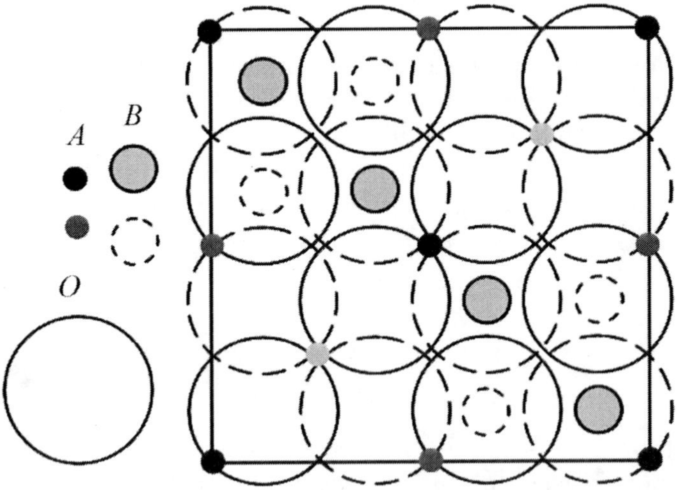

Figure 1. The normal spinel structure of Cobalt Ferrite $CoFe_2O_4(CoO{\cdot}Fe_2O_3)$ with a stoichiometry denoted by AB_2O_4. Only the lower half of the unit cell is shown for projection on (001)

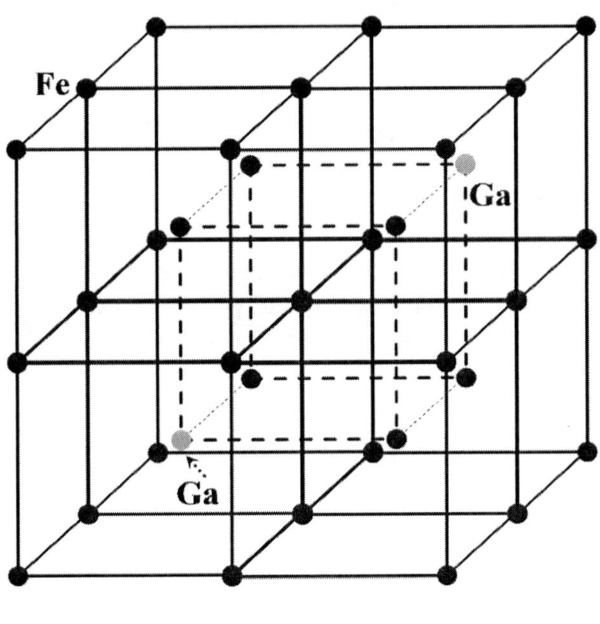

Figure 2. The bcc-structure of the Fe-17Ga alloy (α-Fe)

Terfenol-D can be used in magnetic transducers. It is wellknown that magnetic transducers are increasingly considered as actuators and sensors for numerous aerospace, aeronautic, automotive, industrial, and biomedical applications, for example see [32-34]. Recent progress in studying the cubic structure piezomagnetic crystals Galfenol and Terfenol-D of class m3m for utilization in transducers and sensors can be additionally found in [35-38]. Section 2 theoretically describes SAW propagation in piezomagnetic cubic crystals.

2. Theory

The rectangular coordinate system (x_1, x_2, x_3) is shown in figure 3. The suitable crystal cut and propagation direction was obtained by 45^0-rotation around the x_2-axis directed along the lowest odd-order symmetry axis of a piezomagnetic cubic crystal. Therefore, the piezomagnetic waves with the anti-plane polarization propagate along the crystal surface in direction [101]. It is noted that the x_2-axis is perpendicular to the sagittal plane. Theoretical description of wave propagation in cubic piezomagnetics is similar to that for cubic piezoelectrics, for example see [27,39,40]. The constitutive equations for a piezomagnetic material can be expressed in terms of the strains τ and the magnetic field H. Strains are related to mechanical displacements: $\tau_{ij} = (\partial U_i/\partial x_j + \partial U_j/\partial x_i)/2$ [41]. The governing mechanical equilibrium is $\partial \sigma_{ij}/\partial x_j = 0$ and the governing magnetostatic equilibrium is $\partial B_i/\partial x_i = 0$ where σ_{ij} and B_i are the stress tensor and magnetic flux, respectively. A piezomagnetic medium possesses the elastic C_{ijkl} and piezomagnetic h_{ijk} coefficients as well as the magnetic permeability coefficients μ_{ij} and the medium density r.

It is necessary to write governing equations for the linear case. Constitutive relations are written as follows:

$$\sigma_{ij} = C^{H}_{ijkl}\tau_{kl} - h_{ijm}H_m \tag{1}$$

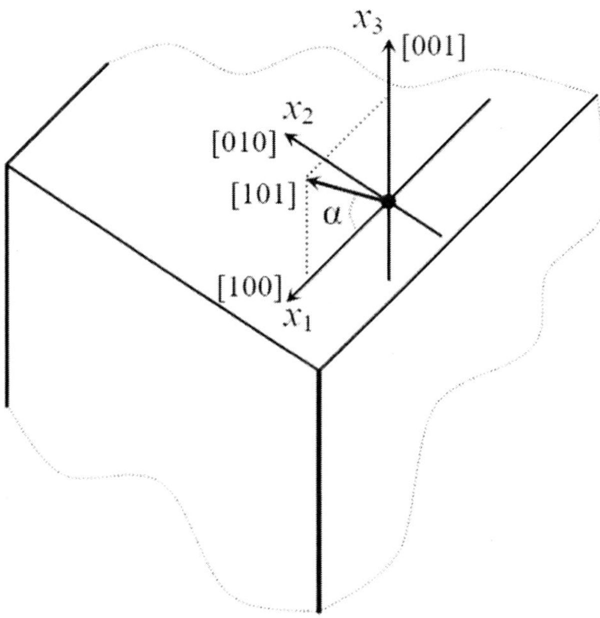

Figure 3. The rectangular coordinate system for a piezomagnetic cubic monocrystal, α = 45°

$$B_m = h_{mij}\tau_{ij} + \mu^{\tau}_{mn}H_n \quad (2)$$

in which σ_{ij} and τ_{ij} are the stress and strain tensors, respectively; B_m and H_m are components of the magnetic induction (i.e. magnetic flux) and magnetic field: $H_m = -\partial\psi/\partial x_m$, where ψ is the magnetic potential. The indices i, j, k, l, m, and n run from 1 to 3. According to Voigt's usual notation, C_{ijkl}, h_{ijm}, and μ_{mn} can be written as 6 × 6, 3 × 6, and 3 × 3 matrices standing for the elastic, piezomagnetic, and magnetic tensors, respectively. They are thermodynamically defined as follows:

$$C^H_{ijkl} = \left(\frac{\partial\sigma_{ij}}{\partial\tau_{kl}}\right)_H \quad (3)$$

$$h_{mij} = h_{ijm} = -\left(\frac{\partial \sigma_{ij}}{\partial H_m}\right)_\tau = \left(\frac{\partial B_m}{\partial \tau_{ij}}\right)_H \qquad (4)$$

$$\mu^\tau_{mn} = \left(\frac{\partial B_m}{\partial H_n}\right)_\tau \qquad (5)$$

where the elastic constants C_{ijkl} are defined at constant magnetic field and the magnetic constants μ_{mn} are defined at constant strain.

In the quasi-static approximation, the equations of motion of an elastic medium and magnetostatics are

$$\frac{\partial \sigma_{ij}}{\partial x_j} = \rho \frac{\partial^2 U_i}{\partial t^2} \quad \text{and} \quad \frac{\partial B_i}{\partial x_j} = 0 \qquad (6)$$

In Equation (6), U_i denote the mechanical displacement components; t is time. Using Equations (1), (2), (6) and $\psi = U_4$, one can write the coupled equations of motion for a piezomagnetic medium in the following form:

$$\rho \frac{\partial^2 U_i}{\partial t^2} = C_{ijkl} \frac{\partial^2 U_l}{\partial x_j \partial x_k} + h_{kij} \frac{\partial^2 \psi}{\partial x_j \partial x_k}$$

$$0 = h_{ijk} \frac{\partial^2 U_k}{\partial x_i \partial x_j} - \mu_{ij} \frac{\partial^2 \psi}{\partial x_i \partial x_j} \qquad (7)$$

Solutions of homogeneous partial differential Equations (7) of the second order are found in the following plane wave form: $U_i = U_i^0 \exp\left[\mathrm{j}(\mathbf{K}_s\mathbf{R} - \omega t)\right]$ and $\psi = \psi^0 \exp\left[\mathrm{j}(\mathbf{K}_s\mathbf{R} - \omega t)\right]$
where the index i runs from 1 to 3. U_i^0 and ψ are initial amplitudes; j $= (-1)^{1/2}$. $k_s r$ denotes the scalar multiplication of two vectors and ω is the angular frequency. $\{k_1, k_2, k_3\} = k\{n_1, n_2, n_3\}$ are the components

of the wavevector k_s, $\{x_1, x_2, x_3\}$ are the components of the real space vector r, and $\{n_1, n_2, n_3\}$ are the directional cosines.

The coupled equations of motion can be readily written in the following simplified form, leaving only equations for waves with polarization perpendicular to the sagittal plane as well as non-zero components of the material tensors for the studied direction [101] of wave propagation:

$$\rho\frac{\partial^2 U_2}{\partial t^2} = C_{44}\left(\frac{\partial^2 U_2}{\partial x_1^2}+\frac{\partial^2 U_2}{\partial x_3^2}\right) - h_{16}\frac{\partial^2 \psi}{\partial x_1^2} + h_{34}\frac{\partial^2 \psi}{\partial x_3^2}$$

$$0 = -h_{16}\frac{\partial^2 U_2}{\partial x_1^2} + h_{34}\frac{\partial^2 U_2}{\partial x_3^2} - \mu_{11}\left(\frac{\partial^2 \psi}{\partial x_1^2}+\frac{\partial^2 \psi}{\partial x_3^2}\right) \tag{8}$$

In Equation (8), the mechanical displacement component U_2 is directed along the x_2-axis, see figure 3:

$$U_{2,4} = U_{2,4}^{0}\exp\left[jk\left(n_1 x_1 + n_3 x_3 - V_{ph} t\right)\right] \tag{9}$$

where the phase velocity is defined as v_{ph} = w/k (k is the wavenumber in direction of wave propagation). For piezomagnetic cubic crystals, it is possible to write $C_{44} = C_{66} = C$ and $\mu_{11} = \mu_{33} = \mu$. It is noted that many propagation directions perpendicular to direction [010] can exist. It is also possible to cut a cubic crystal in order to study wave propagation in direction [101] with $h_{14} = h_{36} = 0$ and the non-zero piezomagnetic constants $\{-h_{16}, h_{34}\}$ that is shown in Equation (8). The dependence of the normalized piezomagnetic constants on the cubic crystal cuts and propagation directions is shown in figure 4.

Substituting the mechanical displacement U_2 and magnetic potential $\psi = U_4$ of Equation (9) into Equation (8), the equations of motion can readily be written in the wellknown tensor form, using corresponding GL-components in the Green-Christoffel equation, $(GL_{rw} - \delta_{rw}\rho V_{ph})\, U_r = 0$

[39,41]: $GL_{22} = C(1+n_3^2)$, $GL_{24} = GL_{42} = -h(1-n_3^2)$ and $GL_{44} = -\mu(1+n_3^2)$ with $n_3 = k_3/k$. In the GL-equation, r and w run from 1 to 4, δ_{rw} is the Kronecker delta for r < 4 and w < 4, $\delta_{44} = 0$, $U_r = \{U_1, U_2, U_3, \psi\}$. Therefore, the following system of two homogeneous equations for pure SH-waves can be written as:

$$\begin{pmatrix} GL_{22} - C(V_{ph}/V_{t4})^2 & GL_{24} \\ GL_{42} & GL_{44} \end{pmatrix} \begin{pmatrix} U_2^0 \\ \psi^0 \end{pmatrix} = 0 \quad (10)$$

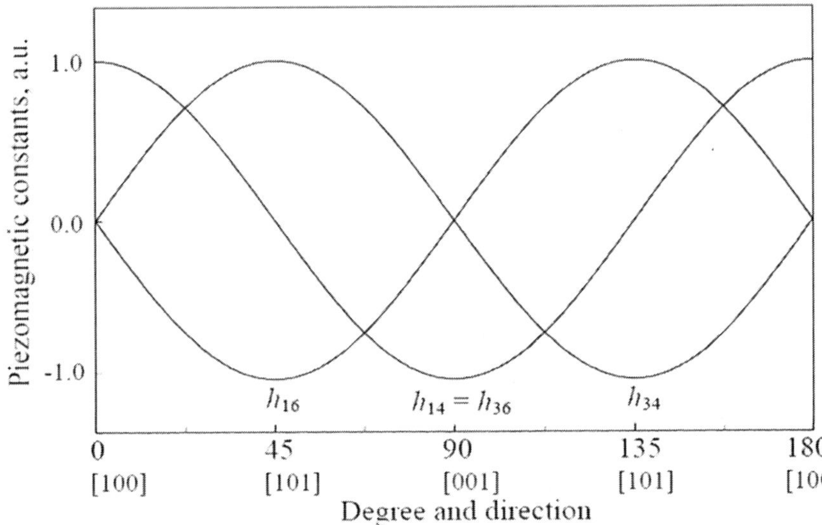

Figure 4. The dependence of the normalized piezomagnetic constants $h_{14}= h_{36}$, h_{16}, and h_{34} on the propagation directions with the Euler angles {0°, θ, 0°} for piezomagnetic cubic crystals, where there is rotation around the x_2-axis

In Equation (10), the directional cosines are defined as follows: $n_1 \equiv 1$, $n_2 \equiv 0$ and $n_3 = n_3$. The velocity V_{t4} is also defined as follows: $V_{t4} = (C_{44}/\rho)^{1/2}$. Setting the matrix determinant equal to zero in Equation (10), the suitable phase velocity V_{ph} satisfying boundary conditions discussed in the following section and four polynomial roots $n_3^{(p)}(V_{ph})$, as well as the functions $U_2^0(V_{ph})$ and $\varphi^0(V_{ph})$, can then be found. For example, the functions can be taken in the following form: $\psi^0 = \mathrm{GL}_{42}$ and $U_2^0 = -\mathrm{GL}_{44}$. It is noted that for the piezomagnetic surface Bleustein-Gulyaev waves, the GL_{24} and GL_{42} components are as follows: $\mathrm{GL}_{24} = \mathrm{GL}_{42} = h\left(1+n_3^2\right)$.

Substituting $m = 1 + n_3^2$ and expanding the matrix determinant in Equation (10), the following polynomial can be obtained from Equation (10):

$$\left(1+K_m^2\right)m_3^2 - Dm_3 + 4K_m^2 = 0 \quad \text{with} \quad D = \left(\frac{V_{ph}}{V_{t4}}\right)^2 + 4K_m^2 \tag{11}$$

of which two roots can be found as follows

$$m_3^{(1,2)} = \frac{D \pm \sqrt{D^2 - 16K_m^2\left(1+K_m^2\right)}}{2\left(1+K_m^2\right)} \tag{12}$$

giving four polynomial roots of Equation (10)

$$n_3^{(1,2,3,4)} = \pm\sqrt{-1+m_3^{(1,2)}} \tag{13}$$

The polynomial roots $n_3^{(1,2,3,4)}$ actually represent the eigenvalues. For each eigenvalue $n_3^{(n)}$, the corresponding eigenvector can be found in the following form: $(U_2^{(n)}, \psi^{(n)})$ where the index n runs from 1 to 4. The coefficient K_m^2 in Equations (11) and (12) is the static coefficient of the magnetomechanical coupling (CMMC):

$$K_m^2 = h_{16}^2/\left(C_{44}\mu_{11}\right) \tag{14}$$

It is also noted that the speed V_{tm} of the bulk SH-wave is given by the following Formula:

$$V_{tm} = V_{t4}\left(1 + K_m^2\right)^{1/2} \tag{15}$$

Analyzing the roots for propagation direction [101] in Equations (12) and (13), it can be found that all complex roots will be calculated when the expression under the square root in Equation (12) is negative. That fulfills for velocities V_{ph} being lower than some velocity V_{Km} obtained solving the following Equation from (12):

$$D^2 - 16K_m^2\left(1 + K_m^2\right) = 0 \tag{16}$$

and defined by the following formula:

$$V_{Km} = A_{Km}V_{t4} \text{ with } A_{Km} = 2\sqrt{K_m\sqrt{1 + K_m^2} - K_m^2} \tag{17}$$

It is clearly seen in Equation (17) that the factor A_{Km} shown in figure 5 is a function of the CMMC K_m^2. The other function f $(K_m^2) = (1 + K_m^2)^{1/2} = V_{tm}/V_{t4}$ from Equation (15) is also shown in figure 5. The function A_{km} $(K_m^2 = K_0^2 = 1/3)$ approaches the function f $(K_m^2) = (1 + K_m^2)^{1/2}$ giving the following equality: $V_{Km} = V_{tm}$ (see figure 5). It is noted that only complex polynomial roots can exist when $V_{ph} < V_{Km}$. The CMMC $K_0^2 = 1/3$ is readily found by substituting the velocity V_{tm} from Equation (15) instead of the phase velocity V_{ph} in Equation (16). Note that when $K_m^2 < K_0^2$ there are all imaginary roots for $V_{ph} > V_{Km}$, but a big $K_m^2 > K_0^2$ gives real roots when $V_{ph} > V_{Km}$. It is also noted that only complex or imaginary roots with negative imaginary parts are chosen in order to cope with wave damping towards the depth of a crystal corresponding to negative values of the x_3-axis shown in figure 3.

Table 1 lists experimental data for Galfenol (83% of Fe and 17% of Ga) obtained in [42,43]. It is clearly seen that the static CMMC K_m^2 for cubic Galfenol can be as high as~0.42 that is larger than 1/3, see also table 2. The Galfenol density ρ was taken from [31], which can depend on material synthesis. For example, [31] gives a Galfenol density of ~5900 kg/m³ for a typical sintered disk that is only 74% of maximum possible density. It is noted that the material density of Fe is ~7848 kg/m³(see [44]). According to [31,42] as well as many

Galfenol Workshops at the University of Maryland, the shear elastic constant c_{44} = c for Galfenol relatively slightly depends on both temperature and concentration of ga in fe. The elastic constant c can have values from ~120 GPa to ~135 gPa. According to [31], the static CMMC K_m^2 for cubic Terfenol-D can be usually about two times larger than that for Galfenol under the same conditions. However, it is interesting to compare some characteristics of Galfenol and Terfenol-D. Therefore, Terfenol-D with a very big K_m^2 ~1.2 was chosen from [45] for numerical calculations suggesting that Terfenol-d has the cubic symmetry. The piezomagnetics $CoFe_2O_4$ [16,46] is also listed in table 2 for comparison, suggesting the cubic symmetry for the crystal.

Since the experimental data listed in **Table 1** is given for different experimental conditions, it is possible to calculate the single piezomagnetic constant h = h_{14} for cubic Galfenol, using known values of K_m and C, as well as the measured values of magnetic permeability μ given in table 1. The strain τ and magnetic flux B can be also written based on the following linear constitutive piezomagnetic equations:

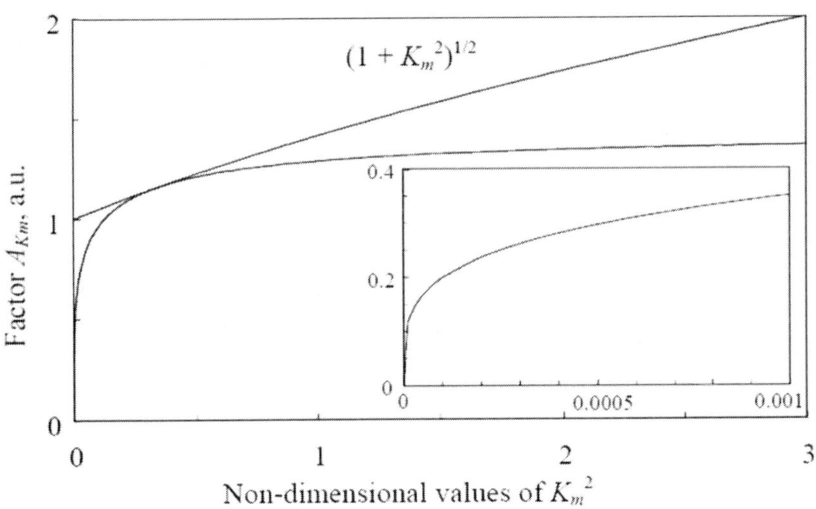

Figure 5. The dependence of both the factor A_{Km} and the function $(1 + K_m^2)^{1/2}$ on the values of K_m^2 from 0 up to 3. The insertion shows the $A_{Km}(K_m^2)$ for small values of the K_m^2

Table 1. The material characteristics following [42,43] for Galfenol $Fe_{0.83}Ga_{0.17}$ (m3m); $E^B = 7.7 \times 10^{10}$ [N/m²]. The magnetic permeability coefficient of a vacuum is $\mu_0 = 4\pi \times 10^{-7}$ [H/m]~12.566371×10^{-7} [(V × s)/(A × m)]. The static coefficient of magnetomechanical coupling (CMMC) K_m in the last column was also announced in [42,43]

Stress Ξ, 10^6 [N/m²]	q, 10^{-9} [m/A]	μ^σ/μ_0,	K_m
16.0	58	285	0.65
27.5	34	174	0.54
50.5	22	108	0.47
85.0	16	70	0.42

Table 2. The elastic C_{44} and piezomagnetic h_{14} constants as well as the material density ρ and magnetic permeability coefficient μ_{11} for the piezomagnetic cubic crystals: Galfenol [42,43], $CoFe_2O_4$ [16,46], and Terfenol-D [45]. The coefficient of magnetomechanical coupling (CMMC) K_m^2 in the last column can be calculated with Formula (14)

Material (symmetry group)	Density, ρ [kg/m³]	C_{44}, 10^{10} [N/m²]	h_{14} [N/(A·m)]	$\mu_{11} = \mu^t$, 10^{-6} [N/A²]	K_m^2
Galfenol (m3m)	7973	12.70	3331.34	206.83	0.4225
Galfenol (m3m)	7973	12.70	2395.05	154.90	0.2916
Galfenol (m3m)	7973	12.70	1722.32	105.74	0.2209
Galfenol (m3m)	7973	12.70	1273.98	72.45	0.1764
$CoFe_2O_4$ (m3m)	5300	4.53	550.0	157.00	0.0425
Terfenol-D (m3m)	9250	0.60	167.7	3.97	1.1807

$$\tau = S^H \sigma + qH \tag{18}$$

$$B = q^* \sigma + \mu^\sigma H \tag{19}$$

where s^H is the compliance at constant magnetic field H and μ^σ is the magnetic permeability at a constant stress σ. Note that Equation (19) incorporates the direct magnetostrictive effect in which magnetic flux is generated by stresses in the piezomagnetic material. The piezomagnetic constants q and q* are thermodynamically defined as follows, assuming q = q*:

$$q = \left(\partial\tau/\partial H\right)_\sigma \text{ and } q^* = \left(\partial B/\partial\sigma\right)_H \tag{20}$$

The piezomagnetic coefficient q symbolizes the axial magnetostriction coefficient and represents the change in strain per change in magnetic field at a constant stress. The parameter q* represents the change in magnetic induction per change in stress at a constant magnetic field.

The short circuit elastic modulus E^B represents the stiffest-material condition, which occurs when all available magnetic energy has been transduced into elastic potential energy. When energy is transferred from the elastic to the magnetic regime the effective modulus decreases to the value of E^H. It has been shown by invoking linear transduction considerations that these two elastic moduli are related by the magnetomechanical coupling coefficient K_m^2 ($0 \leq K_m^2 \leq 1$) as follows [47]:

$$E^H = E^B\left(1 - K_m^2\right) \tag{21}$$

In a similar manner, the intrinsic or uncoupled magnetic permeability μ^σ of Equation (19) can be reduced to a value corresponding to the constant strain permeability μ^ε because of the energy conversion from the magnetic to the elastic regime. The permeabilities are related by:

$$\mu^\varepsilon = \mu^\sigma\left(1 - K_m^2\right) \tag{22}$$

It is emphasized that expressions (21) and (22) are only valid for linear systems and significant corrections are necessary in order to extend these formulae to the full nonlinear regime. However, they are highly useful to illustrate the exchange mechanisms taking place in magnetostrictive transduction.

The static CMMC K_m^2 can be also written as follows:

$$K_m^2 = q^2 \Big/ \left(S^H \mu^\sigma \right) \tag{23}$$

Therefore, the single non-zero piezomagnetic constant h can be calculated with the following formula, using Equations (14), (22), and (23):

$$h = K_m \sqrt{C \mu^\sigma \left(1 - K_m^2\right)} \tag{24}$$

Table 2 lists the piezomagnetic constant h for cubic Galfenol corresponding to the experimentally measured values of the magnetic permeability μ^σ at constant stress σ.

For the free space, Laplace's equation of type $D\psi = 0$ is written in the following form: $(k_1^2 + k_3^2)\ \psi_0 = 0$ where μ_0 is the magnetic constant for the free space. The magnetic potential for the free space can be written as follows: $\psi_0 = F^{(0)} \exp(-k_1 x_3) \exp[j(k_1 x_1 - wt)]$. The potential ψ_0 must decrease according to any increase in the coordinate x_3 (see figure 3). It is now possible to use the material constants in mechanical and magnetic boundary conditions.

3. Boundary Conditions for Surface SH-waves

The mechanical boundary conditions of a magnetoelastic material which occupies the half-space $x_3 < 0$ must be satisfied together with the magnetic conditions of a magnetically closed ($B_3 = 0$) or magnetically open ($\psi = 0$) surface. The realization of each of these boundary conditions is described in [17]. There is the single mechanical boundary condition for the normal component of the stress tensor: $\sigma_{32} = 0$ at $x_3 = 0$, where,

$$\sigma_{32} = \sum_{p=1,2} F^{(p)}\left[Ck_3^{(p)}U_2^{(p)} + hk_3^{(p)}U_4^{(p)}\right] \tag{25}$$

Also, there are the magnetic boundary conditions: continuity of the normal component B_3 of the magnetic flux at $x_3 = 0$ being the interface between a vacuum (B_3^f) and the crystal surface ($B_3 = B_3^f$) where,

$$B_3 = \sum_{p=1,2} F^{(p)}\left[hk_3^{(p)}U_2^{(p)} - \mu k_3^{(p)}U_4^{(p)}\right]$$

$$B_3^f = -F^{(0)}\psi_0^f \, \mathrm{j}k_1\mu_0 \tag{26}$$

and continuity of the magnetic potential $U_4 = \psi$ at $x_3 = 0$ ($\psi = \psi^f$) where

$$\psi = \sum_{p=1,2} F^{(p)}\psi^{(p)} \quad \text{and} \quad \psi^f = F^{(0)}\psi_0^f \tag{27}$$

Therefore, using Equations (25)-(27), two homogeneous equations for the case of a magnetically closed surface can be readily written as follows:

$$\begin{pmatrix} Ck_3^{(1)}U_2^{(1)} + hk_3^{(1)}U_4^{(1)} & Ck_3^{(2)}U_2^{(2)} + hk_3^{(2)}U_4^{(2)} \\ \mathrm{j}\mu_0 k_1 U_4^{(1)} & \mathrm{j}\mu_0 k_1 U_4^{(2)} \end{pmatrix} \begin{pmatrix} F^{(1)} \\ F^{(2)} \end{pmatrix} = 0 \tag{28}$$

For a magnetically open surface, the following homogeneous equations can also be written in the corresponding form:

$$\begin{pmatrix} Ck_3^{(1)}U_2^{(1)} + hk_3^{(1)}U_4^{(1)} & Ck_3^{(2)}U_2^{(2)} + hk_3^{(2)}U_4^{(2)} \\ hk_3^{(1)}U_2^{(1)} - (\mu k_3^{(1)} - \mathrm{j}\mu_0 k_1)U_4^{(1)} & hk_3^{(2)}U_2^{(2)} - (\mu k_3^{(2)} - \mathrm{j}\mu_0 k_1)U_4^{(2)} \end{pmatrix} \begin{pmatrix} F^{(1)} \\ F^{(2)} \end{pmatrix} = 0 \tag{29}$$

Therefore, the corresponding boundary-condition determinants (BCDs) of matrices in Equations (28) and (29) for both cases represent complex numbers.

The complete mechanical displacement U_2^{Σ} and magnetic potential $\psi^{\Sigma} = U_4^{\Sigma}$ can be written in the plane wave form as follows:

$$U_{2,4}^{\Sigma} = \sum_{p=1,2} F^{(p)} U_{2,4}^{0(p)} \exp\left[jk\left(n_1 x_1 + n_3^{(p)} x_3 - V_{ph} t\right)\right] \tag{30}$$

The corresponding weight functions $F^{(1)}$ and $F^{(2)}$ are found from Equations (28) and (29), which can give the same eigenvectors ($U_2^{0(1)}$, $\psi^{0(1)}$) and ($U_2^{0(2)}$, $\psi^{0(2)}$) for two equal eigenvalues $n_3^{(1)} = n_3^{(2)}$, and hence $F^{(1)} = -F^{(2)}$. It is obvious that for this case the weight factors $F^{(1)} = -F^{(2)}$ will zero the complete mechanical displacement U_2^{Σ} and magnetic potential ψ^{Σ} in Equations (28) and (29) giving "latent" characteristics in Equation (30). On the other hand, unequal eigenvalues $n_3^{(1)}$and $n_3^{(2)}$ give different eigenvectors ($U_2^{0(1)}$, $\psi^{0(1)}$) and ($U_2^{0(2)}$, $\psi^{0(2)}$). It is thought that the V_{ph}-solutions with the "latent" characteristics must be experimentally verified due to the fact that some elements of crystals symmetry (screw axis or glide reflection) can be broken near the surface. The same relates to simple reflections and axis if the surface restricts the crystal in such a way that it breaks some simple elements of the crystal class group. It is also noted that V_{ph}- solutions cannot be found when the surface BleusteinGulyaev waves are studied that is an additional difference for finding the BG-waves and USZW-waves.

4. Results and Discussions

Acoustic wave propagation along the surface of piezoelectric material in direction [101] as an instability problem can sustain the surface SH-waves in cubic piezoelectrics that was recently discovered in [27]. Concerning piezomagnetics, it is thought that piezomagnetic SH-SAW can also propagate in direction [101] in cubic piezomagnetics, for instance, crystals of class m3m. Indeed, all cubic crystals of the class possess no piezoelectricity according to [48,49]. Note that concerning transversely-isotropic materials, solutions for the surface BG-waves in piezomagnetics can be written in an identical manner to the wave solutions in piezoelectrics [50]. It is also noted that the velocity V_{BGmo} of piezomagnetic surface BG-wave propagating along the magnetically open surface of transversely-isotropic piezomagnetics (treating the simplest case) can be found with the following well-known formula:

$$V_{BGmo} = V_{tm}\left(1-\left(\frac{K_m^2}{\left(1+K_m^2\right)\left(1+\mu/\mu_0\right)}\right)^2\right)^{1/2} \quad (31)$$

The velocity V_{BGmc} of piezomagnetic BG-wave with the magnetically closed surface can be calculated with the following formula:

$$V_{BGmc} = V_{tm}\left(1-\left(\frac{K_m^2}{1+K_m^2}\right)^2\right)^{1/2} \quad (32)$$

Note that Formula (31) for the velocity V_{BGmo} is not applicable for the case of cubic piezomagnetics. For example, formula (31) gives $V_{BGmo}\sim 4760.093893$ m/s for Galfenol with the largest value of $K_m^2 > 1/3$ in the first row in table 3. This value of V_{BGmo} is close to that of V_{tm}, but not to the value of V_{Km} (see table 3) representing the right result for cubic piezomagnetics with $K_m^2 > 1/3$. The same there is for Terfenol-D with $K_m^2\sim 1$ in table 2, for which the incorrect velocity calculated with Formula (31) is $V_{BGmo}\sim 1179.199962$ m/s.

For cubic piezomagnetics with $K_m^2 < 1/3$, Formula (31) also gives values of V_{BGmo} being closer to corresponding values of V_{tm} than true values of $V_{USZW,o}$. For example, the velocity $V_{USZW,o}$ for Galfenol in the fourth row in table 3 with the smallest value of $K_m^2 < 1/3$ (see also table 2) is $V_{USZW,o}\sim 4328.763803$ m/s. However, Formula (31)

Table 3. The velocities v_{Km}, V_{t4}, and V_{tm} (all in m/s) as well as the velocities $V_{USZW,o}$ and $V_{USZW,c}$ of the ultrasonic surface Zakharenko waves for [101] propagation direction in the piezomagnetic cubic crystals. The last column gives values of the K^2 calculated with Formula (33). All the Galfenols correspond to those listed in tables 1 and 2

Material (symmetry group)	Structure	V_{Km}	V_{t4}	V_{tm}	$V_{USZW,o}$	$V_{USZW,c}$	K^2, %
Galfenol (m3m)	bcc	4740.796256	3991.085010	4760.101550	4740.750567	4545.299858	8.24
Galfenol (m3m)	bcc	4530.187138	3991.085010	4535.795584	4535.727732	4418.693612	5.16
Galfenol (m3m)	bcc	4360.495418	3991.085010	4409.913035	4409.878451	4337.133082	3.30
Galfenol (m3m)	bcc	4217.255080	3991.085010	4328.798060	4328.763803	4279.858496	2.26
$CoFe_2O_4$ (m3m)	Spinel	2396.906402	2923.554311	2985.080877	2985.080691	2982.595557	0.17
Terfenol-D (m3m)	-	1048.738635	805.3872662	1189.319685	1047.759141	999.920958	9.13

gives V_{BGmo} ~4328.783915 m/s that is closer to V_{tm}~ 4328.798060 m/s. Therefore, the differences $\Delta_{BG} = V_{BGmo} - V_{tm}$ and $\Delta_{USZW} = V_{USZW,o} - V_{tm}$ are ~0.014 m/s and ~0.034 m/s, respectively. For comparison, for the piezomagnetic $CoFe_2O_4$ with the smallest value of K_m^2 ~4% from table 2, the true velocity $V_{USZW,o}$ in table 3 is already very close to the velocity V_{BGmo}~2985.08072 m/s, which is also very close to the velocity V_{tm}. Note that the relative magnetic permeability μ/μ_0 in Formula (31) for piezomagnetics is about one order larger than the corresponding relative dielectric constant $\varepsilon/\varepsilon_0$ for piezoelectrics for the same $K_m^2 = K_e^2$, where $K_e^2$is the static coefficient of electromechanical coupling (CEMC), and ε and ε_0 are the single electric constants for a piezoelectrics and free space, respectively. That can result in mm/s-difference for piezomagnetics compared with cm/s-difference for piezoelectrics between the corresponding BG-wave velocity and bulk wave velocity V_t. That can result in significantly larger penetration depth of surface BG and USZW-waves for piezomagnetics towards negative values of the x_3-axis in figure 3. It is also noted that the velocity V_{BGmc} for a magnetically closed surface can be readily calculated with Formula (32), with which the velocity of interfacial Maerfeld-Tournois waves [12] is also calculated. The interfacial Maerfeld-Tournois waves can propagate along the interface of two transverselyisotropic half-spaces of class 6 mm. Concerning cubic piezomagnetics, it is expected that some interfacial waves can also propagate that can be further researched.

Using the velocities $V_{USZW,o}$ and $V_{USZW,c}$ for the USZWwaves, it is possible to evaluate the coefficient of magnetomechanical coupling (CMMC) K^2 for the piezomagnetic cubic crystals listed in table 3. Indeed, it is thought that it is possible to use the well-known formula that is used for piezoelectrics. Therefore, the coefficient K^2 can be evaluated with the following formula:

$$K^2 = 2\left(\frac{V_{USZW,o} - V_{USZW,c}}{V_{USZW,o}}\right) \tag{33}$$

For Galfenol in tables 2 and 3, the evaluated value of K^2 is as high as 2.26% for the smallest value of K_m^2. The value of K^2 increases with increase in K_m^2 to the biggest value of 8.25%. This value of 8.25% for the piezomagnetic Galfenol in the first row of table 3 can be compared with that for Terfenol-D in the Table: K^2(Terfenol-D)~ 9.13%. It is clearly seen in the second and third tables that a big value of K_m^2~1.2 for Terfenol-D being about three times larger than the biggest value of K_m^2 for Galfenol does not give a significant rise in the coefficient K^2. The rise is smaller than 1%. That can be explained using the following fact: the velocity $V_{USZW,o}$ for cubic piezomagnetics with a giant coefficient K_m^2 is situated slightly below the speed V_{Km}, but not slightly below the speed V_{tm}(see also figure 5). That results in lower coefficient K^2. Therefore, it can be concluded that the coefficient K^2 for cubic piezomagnetics with a giant K_m^2 is about 8% to 9% and cannot be larger than 10%. It is thought that the coefficient K^2 for the surface BG-waves in transversely-isotropic piezomagnetics with a giant K_m^2 can be larger than 10%, according to Formulae (31)–(33) and figure 5. The sample behaviors of the boundarycondition determinant (BCD3) around the phase velocity solutions for the piezomagnetic monocrystals $CoFe_2O_4$ and Galfenol are shown in figures 6 and 7, respectively. **Figure 8** shows the dependence of the V_{ph} on the coefficient K_m for the Fe-17Ga alloy. Note that any change in the material density will change both the velocities V_{t4} and $V_{tm}(V_{t4})$, but not the coefficients K_m^2 and K^2.

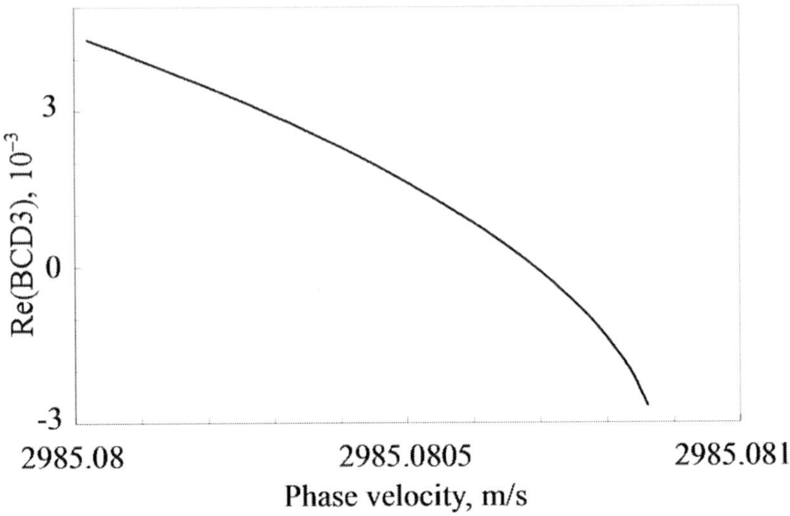

Figure 6. The behavior of the boundary-condition determinant (BCD3) for the piezomagnetic monocrystal $CoFe_2O_4$

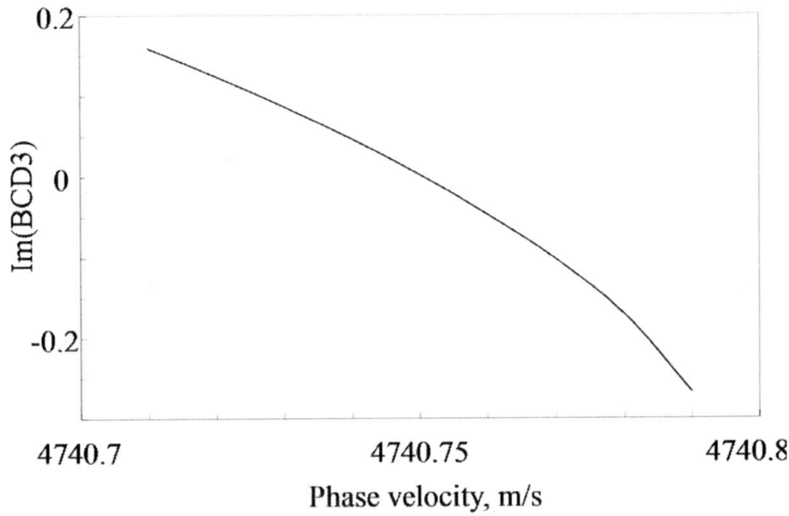

Figure 7. The behavior of the boundary-condition determinant (BCD3) for the piezomagnetic bcc-cubic monocrystal of the Fe-17Ga alloy for $K_m = 0.65$

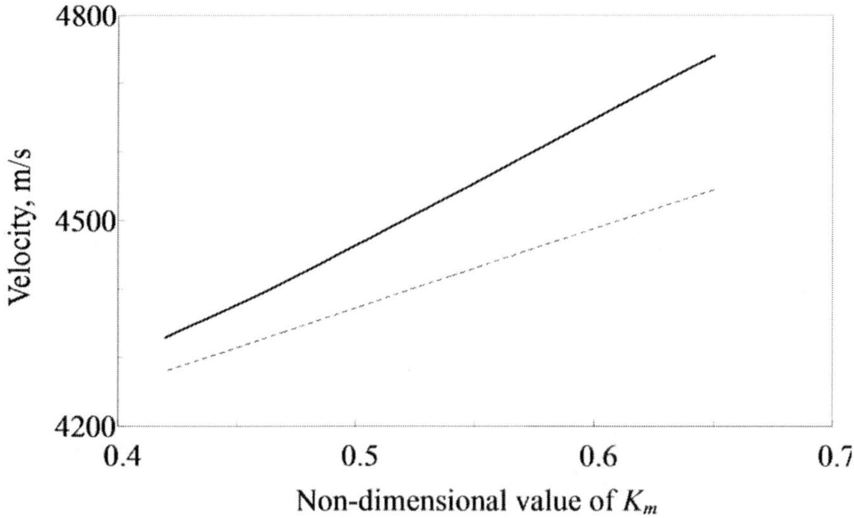

Figure 8. The dependence of the phase velocity V_{ph} on the coefficient $K_m$for the Fe-17Ga alloy, where the functions $V_{USZW,o}(k_m)$ and $V_{USZW,c}(k_m)$ are shown by solid and dashed lines, respectively

5. Conclusions

This paper addresses the theoretical investigations of piezomagnetic acoustic waves propagating along the surface of cubic crystals. The ultrasonic surface Zakharenko waves (USZWs) can propagate in cubic piezomagnetics when direction [101] is parallel to the crystal surface. The computer simulation was done for the widelycalled piezomagnetic crystals Galfenol, Terfenol-D, and $CoFe_2O_4$ suggesting the cubic symmetry. The cubic piezomagnetics Galfenol and Terfenol-D can possess large values of K_m^2, where K_m^2 is the static coefficient of magnetomechanical coupling. The simulations have shown that for a cubic crystal with a large $K_m^2 > 1/3$ (even $K_m^2 > 1$ for Terfenol-D) the coefficient K^2 is as big as 8%-9% and not larger than 10%. It is noted that a big $K_m^2 > 1$ does not give a significant rise to the coefficient K^2 for the case of $K_m^2 \sim 1/3$. That is due to the fact that velocity solutions for large values of $K_m^2 > 1/3$ are found just below the velocity V_{Km}, but not just below the velocity V_{tm}. This theoretical study of cubic piezomagnetics allows use of suitable surfaces of piezomagnetic cubic crystals in non-destructive testing and evaluation that was also recently done for cubic piezoelectrics.

6. Acknowledgements

The author is grateful to all the Referees for their constructive comments.

REFERENCES

1. J. V. Suchtelen, "Product Properties: A New Application of Composite Materials," Philips Research Reports, Vol. 27, No. 1, 1972, pp. 28-37.

2. J. V. D. Boomgaard, D. R. Terrell, R. A .J. Born and H. F. J. I. Giller, "An in Situ Grown Eutectic Magnetoelectric Composite-Material: 1. Composition and Unidirectional Solidification," Journal of Materials Science, Vol. 9, No. 10, 1974, pp. 1705-1709.

3. A. M. J. G. V. Run, D. R. Terrell and J. H. Scholing, "An in Situ Grown Eutectic Magnetoelectric Composite-Material: 2. Physical Properties," Journal of Materials Science, Vol. 9, No. 10, 1974, pp. 1710-1714.

4. J. V. D. Boomgaard, A. M. J. G. V. Run and J. V. Suchtelen, "Piezoelectric-Piezomagnetic Composites with Magnetoelectric Effect," Ferroelectrics, Vol. 14, 1976, pp. 727-728.

5. V. E. Wood and A. E. Austin, "Magnetoelectric Interaction Phenomena in Crystals," In A. J. Freeman, H. Schmid, Eds., Gordon and Breach Science Publishers, Newark, New York, 1975, pp. 181-194.

6. Y. N. Venevtsev, V. V. Gagulin and I. D. Zhitomirsky, "Material Science Aspects of Seignette-Magnetism Problem," Ferroelectrics, Vol. 73, 1987, pp. 221-248.

7. M. Fiebig, "Revival of the Magnetoelectric Effect," Journal of Physics D: Applied Physics, Vol. 38, No. 8, 2005, pp. 123-152.

8. P. Curie, "Sur la Symétrie des Phénomènes Physiques: Symétrie d'un Champ électrique et d'un Champ Magnétique," Journal de Physique, 3e Série, Vol. 3, 1894, pp. 393-415.

9. L. Rayleigh, "On Waves Propagated along the Plane Surface of Elastic Solids," Proceedings of the London Mathematical Society, Oxford, Vol. 17, 1885, pp. 4-11.

10. J. L. Bleustein, "A New Surface Wave in Piezoelectric Materials," Applied Physics Letters, Vol. 13, No. 12, 1986, pp. 412-413.

11. Y. V. Gulyaev, "Electroacoustic Surface Waves in Solids," Soviet Physics Journal of Experimental and Theoretical Physics Letters, Vol. 9, 1969, pp. 37-38.

12. C. Maerfeld and P. Tournois, "Pure Shear Elastic Surface Wave Guide by the Interface of two Semi-Infinite Media," Applied Physics Letters, Vol. 19, No. 4, 1971, pp. 117-118.

13. E. Danicki, "New Interfacial Shear Wave in Piezoelectrics," Applied Physics Letters, Vol. 64, No. 8, 1994, pp. 969-970.

14. A. E. H. Love, "Some Problems of Geodynamics," Cambridge University Press, London, 1911.

15. C. W. Nan, "Magnetoelectric Effect in Composites of Piezoelectric and Piezomagnetic Phases," Physical Review B, Vol. 50, No. 9, 1994, pp. 6082-6088.

16. S. Srinivas, J. Y. Li, Y. C. Zhou and A. K. Soh, "The Effective Magnetoelectroelastic Moduli of Matrix-Based Multiferroic

Composites," Journal of Applied Physics, Vol. 99, No. 4, 2006, pp. (043905)1-7.

17. V. I. Al'shits and A. N. Darinskii and J. Lothe, "On the Existence of Surface Waves in Half-Infinite Anisotropic Elastic Media with Piezoelectric and Piezomagnetic Properties," Wave Motion, Vol. 16, No. 3, 1992, pp. 265- 283.

18. A.-K. Soh and J.-X. Liu, "Interfacial Shear Horizontal Waves in a Piezoelectric-Piezomagnetic Bi-Material," Philosophical Magazine Letters, Vol. 86, No. 1, 2006, pp. 31-35. [Paper reference:1]

19. K.-Q. Hu and G.-Q. Li, "Constant Moving Crack in a Magnetoelectroelastic Material under Anti-Plane Shear Loading," International Journal of Solids and Structures, Vol. 42, No. 9-10, 2005, pp. 2823-2835. [Paper reference:1]

20. X.-F. Li, "Dynamic Analysis of a Cracked MagnetoelectrOelastic Medium under Anti-Plane Mechanical and InPlane Electric and Magnetic Impacts," International Journal of Solids and Structures, Vol. 42, No. 11-12, 2005, pp. 3185-3205.

21. J. Chen, E. Pan and H. Chen, "Wave Propagation in Magneto-Electro-Elastic Multilayered Plates," International Journal of Solids and Structures, Vol. 44, No. 3-4, 2007, pp. 1073-1085.

22. K. Mori and M. Wuttig, "Magnetoelectric Coupling in Terfenol-D/polyvinylidenedifluoride Composites," Applied Physics Letters, Vol. 81, No. 1, 2002, pp. 100-101.

23. M. I. Bichurin, D. A. Filippov, V. M. Petrov, V. M. Laletsin, N. Paddubnaya and G. Srinivasan, "Resonance Magnetoelectric Effects in Layered Magnetostritive-Piezoelectric Composites," Physical Review B, Vol. 68, No. 13, 2003, pp. (132408)1-4.

24. K. Srinivas, G. Prasad, T. Bhimasankaram and S. V. Suryanarayana, "Electromechanical Coefficients of

Magnetoelectric PZT-$CoFe_2O_4$ Composite," Modern Physics Letters B, Vol. 14, No. 17-18, 2000, pp. 663-674.

25. P. Yang, K. Zhao, Y. Yin, J.G. Wan and J.S. Zhu, "Magnetoelectric Effect in Magnetostrictive/Piezoelectric Laminate Composite Terfenol-D/$LiNbO_3$ [(zxtw)-129°/ 30°]," Applied Physics Letters, Vol. 88, No. 17, 2006, pp. (172903)1-3.

26. Y. V. Gulyaev and F. S. Hickernell, "Acoustoelectronics: History, Present State and New Ideas for a New Era," Acoustical Physics, Vol. 51, No. 1, 2005, pp. 81-88.

27. A. A. Zakharenko, "New Solutions of Shear Waves in Piezoelectric Cubic Crystals," Journal of Zhejiang University SCIENCE, Vol. 8, No. 4, 2007, pp. 669-674.

28. W. Kleber, "An Introduction to Crystallography," VEB Verlag Technik, Berlin, 1970.

29. M. J. Dapino, "Nonlinear and Hysteretic MagnetomechaNical Model for Magnetostrictive Transducers," PhD Thesis, Iowa State University, 1999.

30. A. E. Clark, J. B. Restorff, M. Wun-Fogle, T. A. Lograsso and D. L. Schlagel, "Magnetostrictive Properties of Bodycentered Cubic Fe-Ga and Fe-Ga-Al Alloys," IEEE Transactions on Magnetics, Vol. 36, No. 5, 2000, pp. 3238- 3240.

31. R. A. Kellog, "Development and Modeling of Iron-Gallium Alloys," PhD Thesis, Iowa State University, 2003.

32. M. J. Dapino, "On Magnetostrictive Materials and their Use in Adaptive Structures," International Journal of Structural Engineering and Mechanics, Vol. 17, No. 3-4, 2004, pp. 303-329.

33. M. J. Dapino, F. T. Calkins and A. B. Flatau, "Magnetostrictive Devices," Wiley Encyclopedia of Electrical and Electronics Engineering, In. J. G. Webster, Ed., John Wiley and Sons Inc., New York, Vol. 12, 1999, pp. 278- 305.

34. J. B. Restorff, "Magnetostrictive Materials and Devices," Encyclopedia of Applied Physics, Vol. 9, 1994, pp. 229- 244.

35. R. Chung, R. Weber and D. Jiles, "A Terfenol-D Based Magnetostrictive Diode Laser Magnetometer," IEEE Transactions on Magnetics, Vol. 27, No. 6, 1991, pp. 5358-5243.

36. M. J. Dapino, R. C. Smith and A. B. Flatau, "StructuralMagnetic Strain Model for Magnetostrictive Transducers," IEEE Transactions on Magnetics, Vol. 36, No. 3, 2000, pp. 545-556.

37. P. R. Downey and A. B. Flatau, "Magnetoelastic Bending of Galfenol for Sensor Applications," Journal of Applied Physics, Vol. 97, No. 10, 2005, pp. (10R505)1-3.

38. J. Atulasimha, A. B. Flatau and R. A. Kellogg, "Sensing Behavior of Varied Stoichiometry Single Crystal Fe-Ga," Journal of Intelligent Material Systems and Structures, Vol. 17, No. 2, 2006, pp. 97-105.

39. G. W. Farnell and E. L. Adler, "Elastic Wave Propagation in Thin Layers," Physical Acoustics: Principles and Methods, In. W. P. Mason and R. N. Thurston Eds., Academic Press, New York, Vol. 9, 1972, pp. 35-127.

40. C. Lardat, C. Maerfeld and P. Tournois, "Theory and Performance of Acoustical Dispersive Surface Wave Delay Lines," Proceedings of the IEEE, Vol. 59, No. 3, 1971, pp. 355-368.

41. V. E. Lyamov, "Polarization Effects and Interaction Anisotropy of Acoustic Waves in Crystals," MSU Publishing, Moscow, 1983.

42. A. E. Clark, K. B. Hathaway, M. Wun-Fogle, J. B. Restorff, T. A. Lograsso, V. M. Keppens, G. Petculescu and R. A. Taylor, "Extraordinary Magnetoelasticity and Lattice Softening in Bcc Fe-Ga Alloys," Journal of Applied Physics, Vol. 93, No. 10, 2003, pp. 8621-8623.

43. A. E. Clark, "Structural Fe-based Alloys with High Magnetostriction," Galfenol Workshop, University of Maryland, 29 January 2004.

44. A. A. Zakharenko, "On Cubic Crystal Anisotropy for Waves with Rayleigh-Wave Polarization," Non-Destructive Testing and Evaluation, Vol. 21, No. 2, 2006, pp. 61-77.

45. T. J. C. Liu and C. H. Chue, "On the Singularities in a Bimaterial Magneto-Electro-Elastic Composite Wedge under Anti-Plane Deformation," Composite Structures, Vol. 72, No. 2, 2006, pp. 254-265.

46. F. Ramirez, P. R. Heyliger and E. Pan, "Free Vibration Response of Two-Dimensional Magneto-Electro-Elastic Laminated Plates," Journal of Sound and Vibration, Vol. 292, No. 3-5, 2006, pp. 626-644.

47. A. E. Clark, "Magnetostrictive Rare Earth-Fe2 Compounds," Ferromagnetic Materials, In. E. P. Wohlfarth, Ed., North-Holland Publishing Company, Amsterdam, Vol. 1, Chapter 7, 1980, pp. 531-589.

48. V. I. Al'shits and V. N. Lyubimov, "On Existence of Piezoelectric and Piezomagnetic Properties of Anisotropic Media," Soviet Kristallographia, Moscow, Vol. 35, No. 2, 1990, pp. 483-484.

49. D. Royer and E. Dieulesaint, "Elastic Waves in Solids I," Springer-Verlag, New York, 2000.

50. Y. V. Gulyaev, I. E. Dikshtein and V. G. Shavrov, "Magnetoacoustic Surface Waves in Magnetic Crystals at Reorientation Phase Transitions," Uspekhi Phizicheskikh Nauk, Moscow, Vol. 167, No. 7, 1997, pp. 735-750.

8

CHAPTER

Identification of Material Parameters for the Simulation of Acoustic Absorption of Fouled Sintered Fiber Felts

Nicolas Lippitz [1,*,†,‡], Christopher Blech [2,†,‡], Sabine Langer [2,†] and Joachim Rösler [1,†]

[1]*Institute for Materials, TU Braunschweig, Braunschweig 38106, Germany*

[2]*Institute for Engineering Design, TU Braunschweig, Braunschweig 38106, Germany*

Abstract

As a reaction to the increasing noise pollution, caused by the expansion of airports close to residential areas, porous trailing edges are investigated to reduce the aeroacoustic noise produced by flow around the airframe. Besides mechanical and acoustical investigations of porous materials, the fouling behavior of promising materials is an important aspect to estimate the performance in long-term use. For this study, two sintered fiber felts were selected for a long-term fouling experiment where the development of the flow resistivity and accumulation of dirt was observed. Based on 3D structural characterizations obtained from X-ray tomography of the initial materials, acoustic models (Biot and Johnson–Champoux–Allard) in the frame of the transfer matrix method were applied to the sintered fiber felts. Flow resistivity measurements and the measurements of the absorption coefficient in an impedance tube are the basis for a fouling model for sintered fiber felts. The contribution will conclude with recommendations concerning the modeling of pollution processes of porous materials.

Keywords: X-ray; acoustic modeling; absorption; fouling; sintered fiber felt; biot

1. Introduction

The constant exposure to noise affects physical health and can cause insomnia. Especially in areas close to airports, this has become a serious problem. Due to a rising number of flights, it is about to become even more serious. Facing this problem, the noise reduction of commercial aircrafts has become important in engineering. One approach to reducing the noise of commercial aircrafts is to use a porous trailing edge. Experimental and numerical results show that this reduces the aeroacoustic noise, generated by flow around the wings of aircrafts. This noise is a major part of the noise generated by flow around the airframe and landing gear and dominating during landing [**1**,**2**,**3**,**4**]. The functionality of porous materials for noise reduction depends on the porous structure, characterized by porosity, pore size and pore shape. Mounted on trailing edges of airplane wings, the porous materials are exposed to dirt, which is expected to accumulate in the pores during service. This would change the size and shape of the pores as well as the porosity, and, as a consequence, affect the aeroacoustic performance. As the porous material is carefully chosen to reduce the noise of specific phenomena, the affect of dirt must be known. On this basis, the reduced effectiveness can be estimated.

Promising materials for porous trailing edges that achieved a significant reduction of noise in acoustic wind tunnel experiments are sintered fiber felts [5]. To study the influence of dirt accumulation on the acoustic properties of sintered fiber felts, a selection of two different fiber felts was exposed to a long-termed pollution process, and the porous structures were regularly examined over a total period of 12 months. In this paper, the two extremes (zero and twelve months) are compared. Material parameters obtained from X-ray tomography scans are the basis for the application of acoustic models (Biot and Johnson–Champoux–Allard) in the frame of the transfer matrix method. The models are applied to one of the sintered fiber felts. A gradient based optimization algorithm is used to identify missing parameters that are mainly the viscous and the thermal length of the material.

2. Fouling Experiment

For the fouling experiment, samples of two different sintered fiber felts, differing in thickness and fiber diameter, were mounted on the roof rack of a car and fouled for a period of up to 12 months. The samples were separated into a set of samples for flow resistivity measurements, consisting of one sample with a size of 120 mm × 120 mm and a set of three samples for structural inspections with a size of 20 mm × 20 mm for every sintered fiber felt. The flow resistivity was measured after one, six and 12 months and compared to the values of the clean samples. At the same time, the smaller samples were used to check the material for fouling and structural changes. Other than for the flow resistivity measurements, a separate sample was used for the three structural analyses during the experiment. After the last flow resistivity measurements, small circular samples were punched out of the bigger samples used for flow resistivity measurements to measure the absorption coefficient using an impedance tube.

3. Material Characterization

3.1. Structural Characterization

The two different sintered fiber felts (SFF) that were chosen were characterized before the fouling experiments using light microscopy, scanning electron microscopy (SEM) and X-ray tomography. Due to the low density of the organic materials accumulating during the fouling process, X-ray tomography cannot be used to visualize fouling of porous metals. Therefore, only SEM was used to analyze the fouling and changes of porous structure during the 12 months experiment. To ensure the conductivity of the fouled samples, they were sputtered with gold.

3.1.1. Initial Material

The sintered fiber felts were received from "GKN Sinter Metals" (GKN Sinter Metal Filters GmbH, Radevormwald, Germany) and consist of a functional layer and a support grid to ensure mechanical stability. The two different fiber felts differ in thickness of the functional layer, whereas the fiber diameter and porosity of the functional layer are similar. The light microscope images in Figure 1 show cross sections of the sintered fiber felts SFF 100 and SFF 150 used for this study. Structural parameters for the functional layers of the clean fiber felts were determined in [6,7] using three-dimensional CT (Computed Tomography) scans and can be found in Table 1. The support grid of both fiber felts is a plain weave with square mesh. The mesh size is the same in the directions of warp and shut wires and indicated as *w*. It was measured from 2D planes extracted from 3D reconstructions (compare Figure 2). As warp and shut wire are very similar in diameter, an averaged wire diameter dw is used for this work. The mesh size *w* and wire diameter dw were used to calculate the porosity ϕw as well as a hydraulic radius Rh of the support grid [8,9,10]:

$$\phi_w = 1 - \frac{\pi}{4} \frac{d_w \sqrt{d_w^2 + (w + d_w)^2}}{(w + d_w)^2} \tag{1}$$

$$R_h = 1 \sqrt{\phi_w (w + d_w)^2} \tag{2}$$

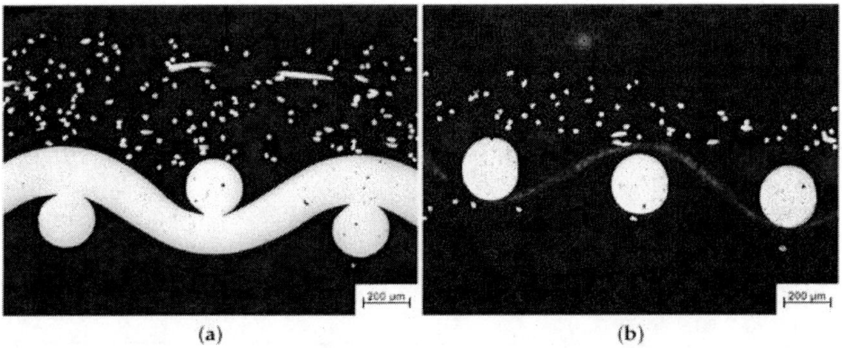

Figure 1. Light microscope images of Sintered Fiber Felts (SFF) before fouling. (**a**) SFF 100; (**b**) SFF 150.

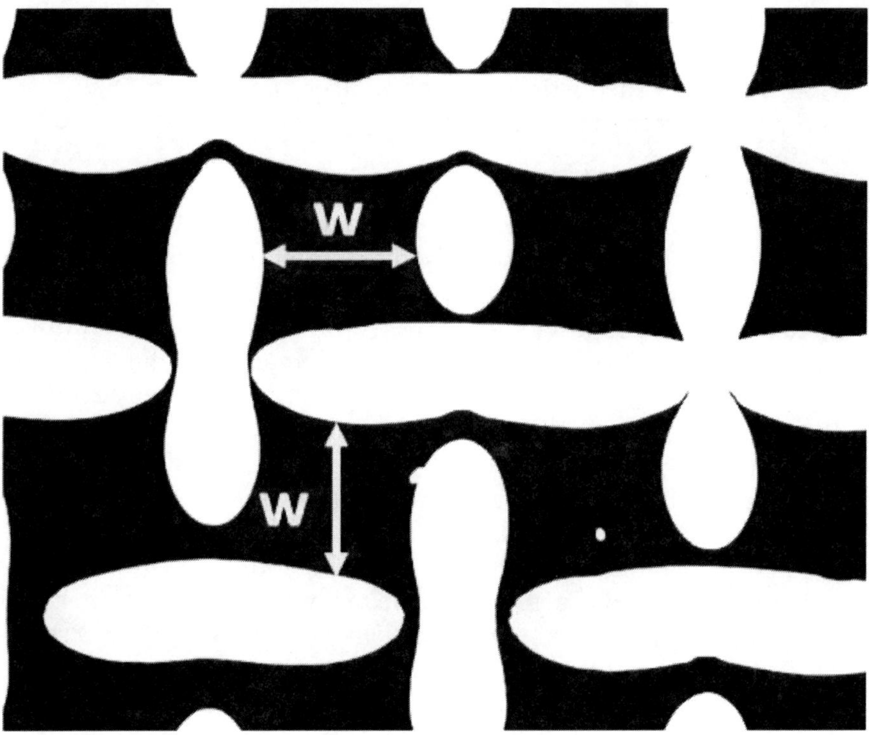

Figure 2. Plane in support grid of SFF 100 extracted from 3D reconstruction.

Table 1. Structural parameters of sintered fiber felts in clean condition. Porosity ϕf of the functional layer, thickness of the functional layers tf, fiber diameter in the functional layer df, wire diameter of the support grid dw, mesh size of the support grid *w*, porosity of the support grid ϕw, hydraulic radius of the support grid Rh.

Material	ϕ_f	d_f (μm)	t_f (μm)	d_w (μm)	w (μm)	ϕ_w	R_h (μm)
SFF 100	0.92	430	26	254	432	0.69	142
SFF 150	0.94	270	24	252	424	0.69	140

The structural parameters of the support grids, measured and calculated, can be found in Table 1. Compared to the functional layer, the pores of the support grid are significantly bigger. Therefore, the acoustic properties are expected to be mainly determined by the functional layer.

3.1.2. Fouled Material

Due to the low density of the dirt, accumulated in the sintered fiber felts, only SEM could be used to examine the fouled samples. Furthermore, the preparation of cross-sectional samples could have altered the fouling pattern through cutting and embedding of the samples. Therefore, only top view images of the functional layer were used. Images of SFF 100 and SFF 150 after 12 months fouling are shown inFigure 3. Both fiber felts show a thin layer of dirt on the fibers. Particles with a size of several micrometers, which could block pores and have a significant impact on the flow resistivity, were not found. Of SFF 100, only the fibers close to the surface are visible in the top view, whereas for SFF 150, which has a significantly thinner functional layer, wires of the support grid are visible through the functional layer. For SFF 100, dirt that blocks pores and influences the flow resistivity could be accumulated deeper in the functional layer, where it cannot be seen from the surface. For SFF 150, a significant impact of pollution in the functional layer on the flow resistivity is not expected.

Figure 3. SEM images of sintered fiber felts after 12 months of fouling. (**a**) SFF 100; (**b**) SFF 150.

3.2. Acoustic Characterization

The flow resistance R and the absorption coefficient α(f) of the specimens are measured before and after the long-term pollution experiment. The flow resistance R is measured using the alternating air flow method as standardised in [11]. Table 2 shows the calculated specific flow resistance RS for clean and fouled sintered fiber felts [11]:

$$R_S = RA \tag{3}$$

where A is the area of the specimen. The flow resistivity $r = R_S/t_f$ (length related flow resistance, also shown in Table 2) is related to the thickness of the functional layer tf. The reason is a significantly lower flow resistivity of the support grid, which is assumed to be negligible for the measurement. As a material parameter, it is suitable to compare sintered fiber felts with different thicknesses and will be used for the following observations.

Table 2. Specific flow resistance, thickness and flow resistivity of clean and fouled SFF 100 and SFF 150.

Parameter	SFF 100		SFF 150	
	Clean	Fouled	Clean	Fouled
R_S (Ns/m^3)	51.26	55.56	15.13	15.48
h_f (m)	0.00043	0.00043	0.00027	0.00027
r (Ns/m^4)	119,208	129,209	56,036	57,333

Due to fouling, the flow resitivity of SFF 100 is increased by 8.4%. In contrast, r of SFF 150 is only slightly increased by 2.3%. This corresponds to the observations made in SEM images of the fouled samples.

A measurement of the support grid only was not possible due to the lower limit of the measurement equipment. Hence, the flow resistivity of the support grid is calculated according to [12], assuming cylindrical pores with circular cross-section and using the hydraulic radius from Table 1:

$$r = \frac{8\eta}{R_h^2 \Phi_w} = 2613\,\mathrm{Ns/m^4} \tag{4}$$

For Rh and Φw, values from Table 1 are used.

The frequency-dependent absorption coefficient α(f) is received applying the standardised method described in [13]. An impedance tube of diameter dI=0.019m is available for the measurement, which results in a lower frequency limit of fl=1140Hz and an upper frequency limit of fu=10,400Hz [13]. This frequency range is of practical relevance and showed a significant reduction of noise in acoustic wind tunnel experiments [2,3,5]. Previous work [14] shows that the applied impedance tube has a sufficient accuracy up to 8000Hz in practice. Hence, the upper limit of investigation is additionally decreased to fu=8000Hz. The functional layer of the SFF is oriented facing the planar acoustic waves. The specimens are punched out of the square cuts of the polluted samples and stapled for the measurement (see Figure 4).

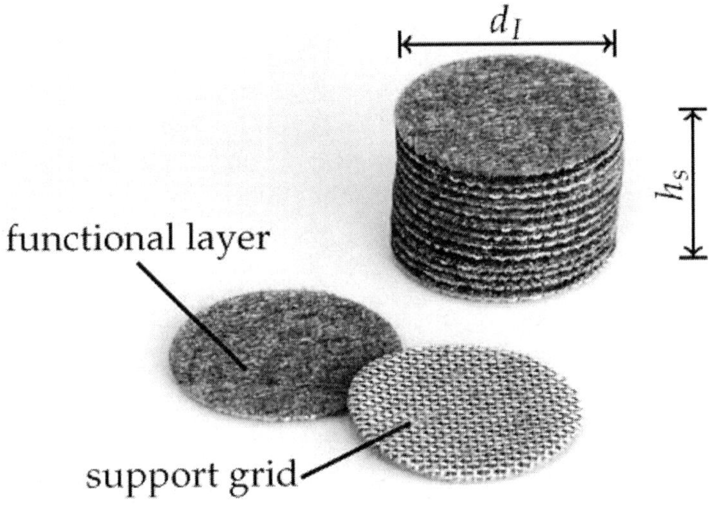

Figure 4. Punched specimens of SFF 100.

Due to the small thickness of 0.0007to0.0009m, a measurement of a single layer is not reasonable. The staple sizes are documented in Table 3. Due to a slight skewness of the specimens and inhomogeneities concerning the thicknesses, the number of specimens does not ideally fit to the staple sizes. Within the experiment, the number of specimens is counted and the total staple size is measured.

Table 3. Staple sizes of punched specimens of SFF 100 and SFF 150.

Material	Staple Size h_s (m)	Number of Specimens N_s (-)
SFF 100	0.0009 0.01 0.02	1 11 21
SFF 150	0.0007 0.01 0.02	1 12 25

The associated results for the frequency-dependent absorption of SFF 100 and SFF 150 coefficient can be seen in Figure 5 and Figure 6. For the measurement of a single layer, both figures point out noise dominating the measurement. A material characterization on that basis is not reasonable.

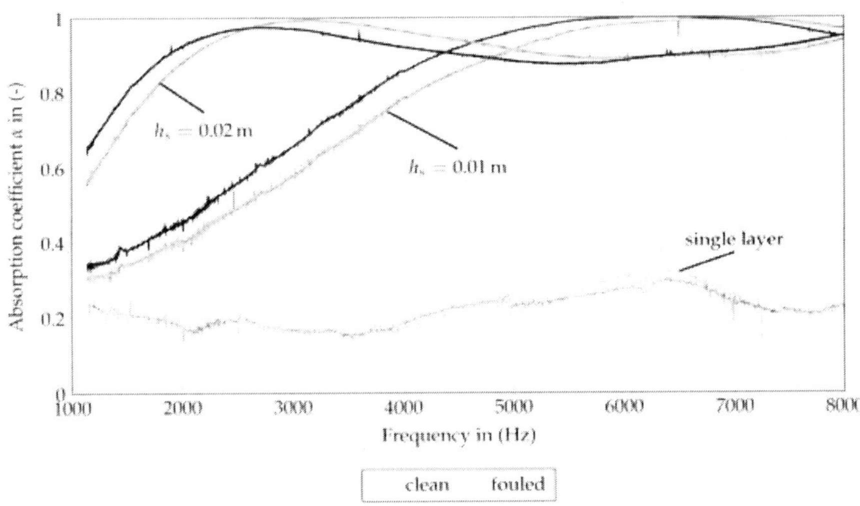

Figure 5. Measured frequency-dependent absorption coefficients of clean (0 M) and fouled (12 M) SFF 100 for one layer and staple sizes of 0.01m and 0.02m.

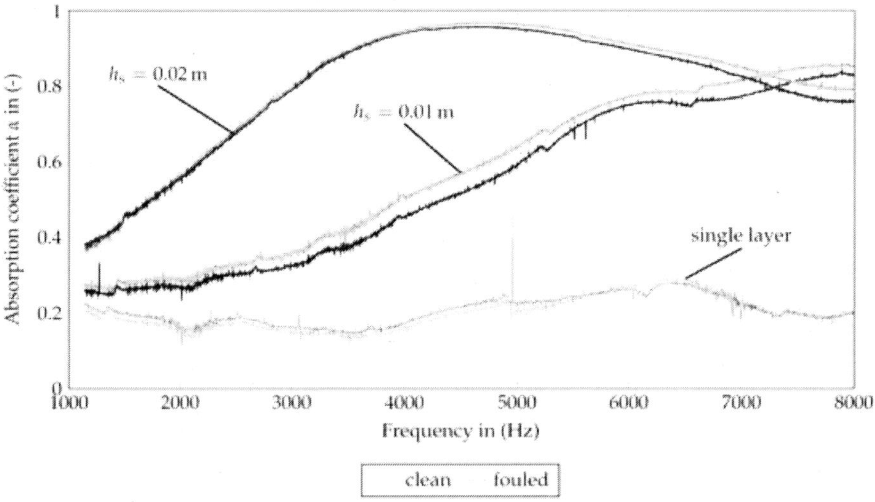

Figure 6. Measured frequency-dependent absorption coefficients of clean (0 M) and fouled (12 M) SFF 150 for one layer and staple sizes of 0.01m and 0.02m.

The SFF 100 (Figure 5) shows different absorption curves for staple sizes hs of 0.01 m and 0.02m. The first maxima with absorption coefficients close to one are located at 7000Hz and 3200Hz, respectively, for hs=0.01 m and 0.02m. The absorption curve of the largest staple size (hs=0.02m) shows the characteristic wave after the fist maximum. In accordance with [12], this zone should be covered to ensure an appropriate parameter identification of the material's characteristic lengths. Hence, this staple size is chosen for the numerical investigation in Section 4.

The absorption curves of the fouled SFF 100 are, for the most part, slightly shifted to lower frequencies and slightly decreased for hs=0.02m. As indicated in Section 3.1, the effect of the fouling process is small.

The SFF 150 (Figure 6) shows a similar behavior to the SFF 100. For the clean SFF 150, the first maximum can only be identified at 4800Hz for hs=0.02m. Consistent with the small change in the flow resistivity, the absorption curve of the fouled SFF 150 is hardly distinguishable from the curves of the clean SFF 150. Concerning the flow resistivity and the absorption, this material seems to be quite resistant to the pollution experiment. Therefore, in the modeling part (Section 4), the focus is laid on the SFF 100.

4. Numerical Investigation

As mentioned in Section 3.2, this part focuses on the numerical investigation of clean and fouled SFF 100. The aim is a modeling of the frequency-dependent absorption coefficient $\alpha(f)$ for normal incidence to inversely receive the material's unknown parameters. The application of these parameters may help to understand fouling processes in more complex porous structures. The unknown parameters are fitted using an in-house implementation of the Transfer Matrix Method (TMM) in combination with a gradient-based optimization algorithm in MATLAB (R2013a, Natick, MA, USA) . The TTM is proposed in [15] as a general method to calculate the response of a infinite extended layered structure excited by a plane wave. For each layer, a transfer matrix is calculated to describe the wave propagation from one face to the other. The following material models are introduced to achieve the layer matrices and form the system matrix. The velocity and pressure of the layered structures are the results. On that basis, the absorption coefficient α of the stapled system can be derived. The in-house implementation is verified including the coupling between different material models by use of results from literature [12].

4.1. Modeling of Stapled Sintered Fiber Felts

The functional layer and the support grid of the SFF 100 material are modeled using different approaches. The response of the functional layer is modeled using Biot's model [16]. This approach considers waves within the structural and fluid phases of the functional layer. As effects in the fluid phase of the support grid on the overall absorption coefficient are expected, the grid is modeled using the Johnson–Champoux–Allard model [17,18]. Here, structural waves are neglected and the grid itself is treated as rigid perforated screen. The parameters needed for the two models are mainly taken from the structural and acoustic characterizations (Section 3.1 and Section 3.2) and given in Table 4.

Table 4. Material parameters used for the acoustic models.

Parameter	Symbol	Unit	Functional Layer	Support Grid
layer thickness	h_S	(m)	0.00043	0.00047
Young's modulus	E	(N/m^2)	$2 \cdot 10^{11}$	-
Poisson ratio	ν	(-)	0.3	-
structural density	ρ_S	(kg/m^3)	8000	-
loss factor	η_l	(-)	0.01	-
fluid density	ρ_F	(kg/m^3)	1.21	1.21
speed of sound	c_F	(m/s)	343	343
porosity	Φ	(-)	0.915	0.691
flow resistivity	r	(Ns/m^4)	119,208	2610
tortuosity	α_∞	(-)	1–1.6	1
viscous length	Λ	(μm)	11 to 123	409
thermal length	Λ'	(μm)	11 to 123	204

Concerning the functional layer, the unknown parameters are the tortuosity $\alpha\infty$, the viscous characteristic length Λ and the thermal characteristic length Λ'.

- **Tortuosity** $\alpha\infty$ is defined by Johnson in 1987 [17]. The value describes the relationship between the squared macroscopic and microscopic velocity to calculate an effective density of the fluid. For common fiber felts, tortuosity is expected near unity [19].

- **Viscous characteristic length Λ** is introduced by Johnson et al. as an extension of the hydraulic radius RH (equal with factor 2 for cylindrical pores) in 1986 [20]. The value is defined as two times the weighted pore's volume integral divided by the weighted pore wall's surface integral. The weighting is realised by the fluid's velocity, respectively, within the volume and on the surface. For common porous materials, the following assumption is taken from literature [21]:

$$\Lambda = \frac{1}{c}\left(\frac{8\alpha_\infty \eta}{R_h \Phi}\right)^{0.5}, \qquad 0.3 \leq c \leq 3.3 \qquad (5)$$

where c is an acoustic value describing the pores and η is the dynamic viscosity of air (η(25∘C)=1.86e−5Pas).

- **Thermal characteristic length** Λ' is defined by Champoux and Allard in 1991 [18] and is used to describe thermal dissipation effects at higher frequencies. Λ' equals two times the pore's volume divided by the pore wall's surface and influences the bulk modulus of the material layer. For common porous materials, the following assumptions are taken from literature [21]:

$$\Lambda' > \Lambda \qquad (6)$$

$$\Lambda' = \frac{1}{c'}\left(\frac{8\alpha_\infty \eta}{R_h \Phi}\right)^{0.5}, \qquad 0.3 \leq c' \leq 3.3 \quad (7)$$

The assumptions considered for the Biot layer modeling the fiber felt lead to the parameters and parameter ranges shown in Table 4. For sintered fiber felts, the characteristic thermal length Λ' is expected to be twice as large as the characteristic viscous length Λ [21].

Concerning the support grid, the unknown parameters are the same three as described above. In the case of the coarse grid (see Figure 1), acoustic waves traveling straight through the grid are expected. Hence, the tortuosity is set equal to one. Furthermore, as the grid is assumed to be a macroscopic regularly patterned fibrous material, the following formula are considered for the characteristic lengths [21]:

$$\Lambda = \frac{1}{2\pi L R_h} \qquad (8)$$

$$\Lambda' = 2\Lambda \quad (9)$$

Again, the corrected value for Rh is applied to receive the characteristic lengths of the support grid, which are also shown in Table 4.

The two modeling approaches are applied alternating to realize the staples (using TMM) documented in Table 3. This results in a total number of layers Ntotal=2∗Ns. Figure 7 shows a schematic sketch of the alternating layers, finished by a hard wall (total reflection).

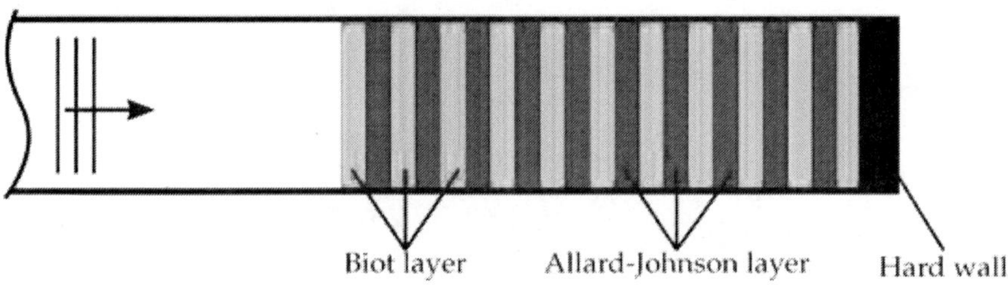

Figure 7. Modeling approach for the acoustic response of stapled specimens.

4.2. Modeling of Fouling

The fouling process of sintered fiber felts leads to dirt located on the fiber's surface, which may influence the materials' macroscopic behavior. Due to small particle sizes of the observed dirt, a need for different modeling approaches is not expected. A modification of the model parameters with regard to the fouling process is described as follows.

In Section 3.1, dirt is mainly identified within the functional layer's microstructure. Hence, the parameters for the Johnson–Champoux–Allard model applied to the layer of the support grid are assumed to be equal before and after the fouling process.

The parameters of Biot's model applied to thefunctional layer are expected to change due to the fouling process. Mechanical parameters of the dirt are not known and the dirt is expected to have no influence on the structural phase of Biot's model. The parameters concerning the fluid phase are modified only. In Table 2, the slightly increased flow resistivity is already shown (8.4%) and can easily be applied for the modeling of the fouled SFF 100. Furthermore, in [22], a pollution parameter is adopted and referred to the flow resistivity:

$$r_{fouled} = \frac{r_{clean}}{\widehat{\Phi}^2} \tag{10}$$

$$\widehat{\Phi} = \sqrt{\frac{r_{clean}}{r_{fouled}}} \tag{11}$$

$$\widehat{\Phi} = 0.96 \tag{12}$$

where $\Phi\hat{}$ describes the relative volume of fluid which is not polluted. In this case, $1-\Phi\hat{}=4\%$ of the fluid volume is taken by dirt. A new porosity for the fouled material can be calculated [22]:

$$\Phi_{fouled} = \widehat{\Phi}\Phi_{clean} = 0.88 \tag{13}$$

The tortuosity $\alpha\infty$ and the characteristic lengths Λ and Λ' are assumed to be equal for clean and fouled sintered fiber felts. The reason is that the observed small particles of dirt should not change these parameters significantly.

4.3. Inverse Parameter Identification

Applying the received parameters (Table 4) for the clean SFF 100 results in a model with three unknown parameters of the Biot layer (functional layer). The following parameters are found using the gradient based optimization algorithm:

$$\alpha_\infty = 1.3 \tag{14}$$

$$\Lambda = 27\,\mu\text{m} \tag{15}$$

$$\Lambda' = 27\,\mu\text{m} \tag{16}$$

As mentioned in Section 4.1, concerning the functional layer, the characteristic thermal length Λ' is expected to be twice as large as the characteristic viscous length Λ [21]. Equality of these parameters is rather expected for straight cylindrical pores [21]. Here, the equality may be explained by the small thickness of the functional layers, each followed by the support grid with significantly larger pores. This may result in a straight transmission of sound waves within the investigated frequency range, similar to cylindrical pores. A tortuosity of 1.3 is near unity, and, therefore, a realistic value for the SFF material. Due to the coupling of 21 double-layers, rotations of the support grids towards each other may result in a slightly increased tortuosity. An additional investigation of separated support grids would help to verify this statement.

Figure 8a shows the comparison of the measured and fitted curve of $\alpha(f)$. In all plots, measurement data is drawn by the use of solid lines and simulation data with dashed ones. Furthermore, the curves of clean SFF are green and the curves of fouled SFF are black. The curves in Figure 8 are not fitting perfectly. Several reasons are assumed to have a major impact on the results. First, the impedance tube has a relatively high inherent absorption. This may result in raised absorption curves in the measurement. Second, the flow resistance is measured using the whole specimen (before punching). Inhomogeneities may lead to the fact that better absorbing parts are located closer to the end of the staple, which may result in higher absorption. Furthermore, including the support grid in the measurement, a slightly raised flow resistance is measured and finally referred to the functional layer. A separation of the layers within the measurement is not possible.

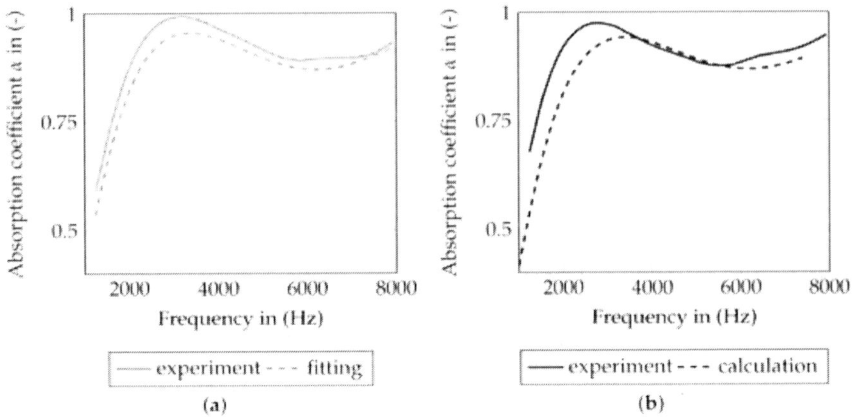

Figure 8. Comparison of numerical and experimental data for a staple size of 0.02m. (**a**) fitting of clean SFF 100; (**b**) calculation (forward solution of the mechanical problem, without any model updating) of fouled SFF 100.

The directly calculated response of the fouled SFF 100 can be seen in Figure 8b. Due to only small changes of r and Φ, the absorption coefficient does not change significantly. The shift of the measured curve to lower frequencies cannot be reproduced properly. Figure 9 additionally shows the difference of the measured absorption curves (a) and the calculated absorption curves (b) applying the fouling model. Only a slight tendency of a decreased absorption coefficient is recognizable in the calculation.

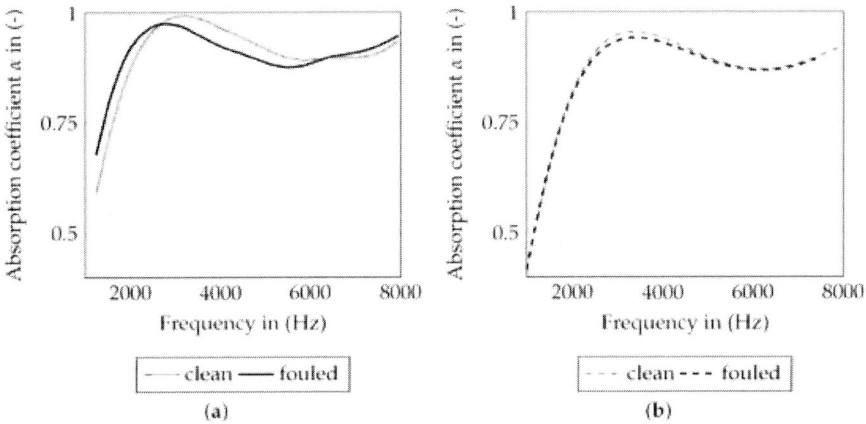

Figure 9. Comparison of clean and fouled SFF 100 for a staple size of 0.02m. (**a**) experimental data; and (**b**) numerical data.

To give first starting points for further investigations of the indicated problems concerning multilayered material configurations with several couplings between small layers and the modeling of fouling, an inverse parameter identification of the fouled material is done. $\alpha\infty$, Λ and Λ' of the Biot layer are fitted using the gradient-based optimization algorithm:

$$\alpha_\infty = 1.5 \quad (17)$$

$$\Lambda = 26\,\mu\text{m} \quad (18)$$

$$\Lambda' = 26\,\mu\text{m} \quad (19)$$

Furthermore, the tortuosity of the support grid was included into the fitting, but the optimizer only changes this value by the amount of 0.2%. Apparently, the introduced degrees of freedom in the functional layer tend to cause an increase of the tortuosity after the pollution process (1.3→1.5). Λ andΛ' are not changed significantly in comparison to the fitting of the clean SFF 100. Figure 10 shows the better fitting curve with the maximum shifted to lower frequencies. Nevertheless, as already stated for the very first fitting of the clean SFF 100, the absorption coefficient is slightly lowered over the complete frequency range.

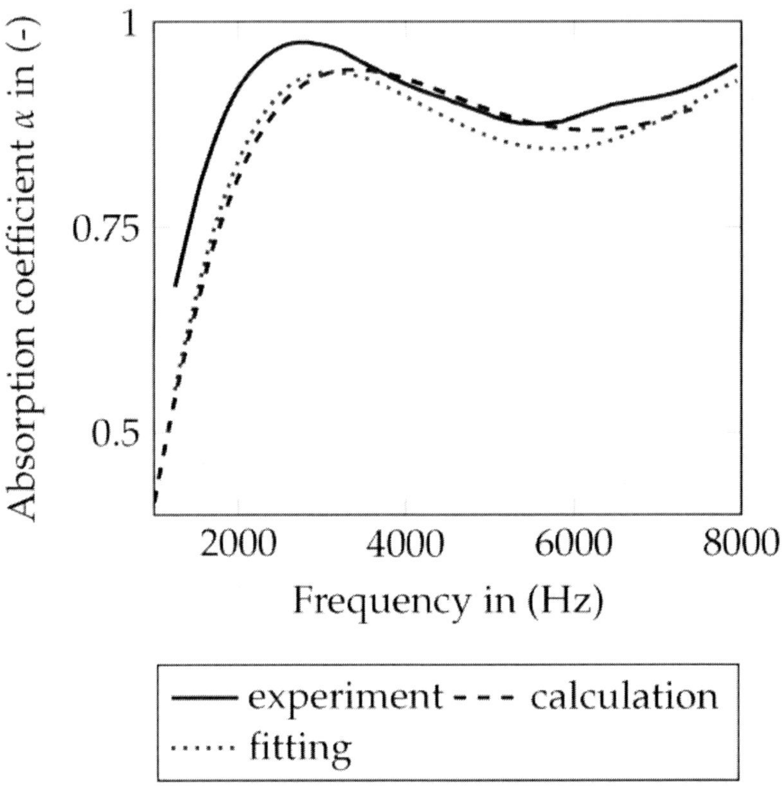

Figure 10. Calculated and fitted curves for fouled SFF 100.

5. Conclusions

In this paper, two types of sintered fiber felts, underlying a fouling process, are investigated concerning the material parameters. In the long-term fouling experiment, the flow resistivity of the sintered fiber felts increases slightly. This corresponds to the rare pollution observed in SEM images. The frequency-dependent absorption coefficient is slightly shifted to lower frequencies and decreased. If this type of porous material is considered in aircraft applications, only small reductions of the performance are expected. Further study must be carried out to ensure a secure statement on the pollution effects in different use cases.

Within a modeling approach on the basis of Biot's theory, the two layers of the SFF 100 material are considered. The multilayered model of the clean material predicts the characteristic shape of the absorption coefficient with a small systematic error. The effect of the fouling process can also be predicted qualitatively applying a simple model. Besides an increase of the flow resistivity and a decrease of the porosity, a changing tortuosity should be considered as well. Further investigation should focus on coupling effects in multilayered material configurations and the development of more detailed modeling of fouling.

Acknowledgments

The authors would like to thank the Sonderforschungsbereich 880 of the Deutsche Forschungsgemeinschaft and its graduate reseach program Modul Graduiertenkolleg (MGK) for the financial support. Special thanks are directed to the DLR (German Aerospace Center) Braunschweig for providing the impedance tube and to the PTB (National Metrology Institute of Germany) Braunschweig for providing the experimental data of the flow resistivity.

Author Contributions

Nicolas Lippitz and Christopher Blech performed the experiments, analyzed the data and wrote the paper; Joachim Rösler and Sabine Langer conceived and designed the experiments.

Conflicts of Interest

The authors declare no conflict of interest. The founding sponsors had no role in the design of the study; in the collection, analyses, or interpretation of data; in the writing of the manuscript, and in the decision to publish the results.

Abbreviations

The following abbreviations are used in this manuscript:

CT	Computed Tomography
SEM	Scanning Electron Microscope
SFF	Sintered Fiber Felt
TMM	Transfer Matrix Method

References

1. Revell, J.; Kuntz, H.; Balena, F.; Horne, C.; Storms, B.; Dougherty, R. Trailing-edge flap noise reduction by porous acoustic treatment. In Proceedings of the 3rd AIAA/CEAS Aeroacoustics Conference, Atlanta, GA, USA, 12–14 May 1997.
2. Herr, M.; Dobrinsky, W. Experimental investigation in low-noise trailing edge design. *AIAA J.* **2005**, *6*, 1167–1175.
3. Herr, M. Design criteria for low-noise trailing-edges. In Proceedings of the 13th AIAA/CEAS Aeroacoustics Conference, Rome, Italy, 21–23 May 2007.
4. Delfs, J.; Fassmann, B.; Lippitz, N.; Lummer, M.; Mößner, M.; Müller, L.; Rurkowska, K.; Uphoff, S. SFB 880: Aeroacoustic research for low noise take-off and landing. *CEAS Aeronaut. J.* **2014**.

5. Herr, M.; Rossignol, K.-S.; Delfs, J.; Mößner, M.; Lippitz, N. Specification of porous materials for low-noise trailing-edge applications. In Proceedings of the 20th AIAA/CEAS Aeroacoustics Conference, Atlanta, GA, USA, 16–20 June 2014.

6. Lippitz, N.; Rösler, J. Potential of metal fiber felts as passive absorbers in absorption silencers.*Metals* **2013**, *3*, 150–158.

7. Lippitz, N.; Hinze, B.; Rösler, J. Characterization of porous materials. In *SFB 880 - Fundamentals of High-Lift for Future Commercial Aircraft, Biennial Report*; TU Braunschweig, Campus Forschungsflughafen: Braunschweig, Germany, 2013; pp. 29–39.

8. Blass, E. Geometrische und strömungstechnische untersuchungen an drahtgeweben. *Chem. Ing. Tech.* **1964**, *7*, 747–758.

9. Armour, J.C.; Cannon, J.N. Fluid through woven screens. *AIChE J.* **2013**, *3*, 415–420.

10. Müller, M.; Piesche, M. Ähnlichkeitsgesetze zur beschreibung des anfangsdruckverlustes metallischer drahtgewebe bei der filtration nicht-newtonscher fluide. *Filtr. Sep.* **2013**, *5*, 284–291.

11. *DIN EN 29053: Akustik. Materialien für Akustische Anwendungen*; Deutsches Institut für Normung: Berlin, Germany, 1993.

12. Atalla, F.; Panneton, R. Inverse acoustical characterization of open cell porous media using impedance tube measurements. *Can. Acoust.* **2005**, *33*, 11–24.

13. *DIN EN ISO 10534-2. Determination of Sound Absorption Coefficient and Acoustic Impedance with the Interferometer*; Deutsches Institut für Normung: Berlin, Germany, 2001.

14. Lippitz, N. *Analyse der Schallabsorptionseigenschaften in Abhängigkeit der Struktur Verschiedener Metallischer Vliese*;

Studienarbeit, TU Braunschweig: Braunschweig, Germany, 2011.

15. Brouard, B.; Lafarge, D.; Allard, J.-F. A general method of modelling sound propagation in layered media. *J. Sound Vib.* **1995**.

16. Biot, M.A. The theory of propagation of elastic waves in a fluid-saturated porous solid. I. Low frequency range. II. Higher frequency range. *J. Acoust. Soc. Am.* **1956**.

17. Johnson, D.L.; Koplik, J.; Dashen, R. Theory of dynamic permeability and tortuosity in fluid-saturated porous media. *J. Fluid Mech.* **1987**, *176*, 379–402.

18. Champoux, Y.; Allard, J.-F. Dynamic tortuosity and bulk modulus in air-saturated porous media.*J. Appl. Phys.* **1991**.

19. Allard, J.-F.; Champoux, Y. New empirical equations for sound absorption in rigid frame fibrous materials. *J. Acoust. Soc. Am.* **1992**.

20. Johnson, D.L.; Koplik, J.; Schwartz, L.M. New pore size parameter characterizing transport in porous media. *Phys. Rev. Lett.* **1986**.

21. Allard, J.-F. *Propagation of Sound in Porous Media: Modeling Sound Absorbing Materials*; Elsevier Applied Science: New York, NY, USA, 1993.

22. Lippitz, N.; Rurkowska, K.; Rösler, J.; Langer, S.C. Fouling behavior of porous metals. *Proc. Mater. Sci.* **2013**.

© 2016 by the authors; licensee MDPI, Basel, Switzerland. This article is an open access article distributed under the terms and conditions of the Creative Commons Attribution (CC-BY) license (http://creativecommons.org/licenses/by/4.0/).

9

CHAPTER

Transverse Crack Detection in 3D Angle Interlock Glass Fibre Composites Using Acoustic Emission

Matthieu Gresil [1,2,*], Mohamed Nasr Saleh [2] and Constantinos Soutis [3]

[1]i-Composites Lab, School of Materials, University of Manchester, 79 Sackville Street, Manchester M1 3NJ, UK

[2]National Composites Certification and Evaluation Facility, University of Manchester, Manchester M1 3NJ, UK

[3]Aerospace Research Institute, University of Manchester, Manchester M1 3NJ, UK

Abstract

In addition to manufacturing cost and production rates, damage resistance has become a major issue for the composites industry. Three-dimensional (3D) woven composites have superior through-thickness properties compared to two-dimensional (2D) laminates, for example, improved impact damage resistance, high interlaminar fracture toughness and reduced notch sensitivity. The performance of 3D woven preforms is dependent on the fabric architecture, which is determined by the binding pattern. For this study, angle interlock (AI) structures with through-thickness binding were manufactured. The AI cracking simulation shows that the transverse component is the one that leads to transverse matrix cracking in the weft yarn under tensile loading. Monitoring of acoustic emission (AE) during mechanical loading is an effective tool in the study of damage processes in glass fiber-reinforced composites. Tests were performed with piezoelectric sensors bonded on a tensile specimen acting as passive receivers of AE signals. An experimental data has been generated which was useful to validate the multi-physics finite element method (MP-FEM), providing insight into the damage behaviour of novel 3D AI glass fibre composites. MP-FEM and experimental data showed that transverse crack generated a predominant flexural mode A0 and also a less energetic extensional mode S0.

Keywords: acoustic emission; Lamb waves; multi-physics finite element; piezoelectric sensors; 3D woven composite materials; structural health monitoring; transverse cracking

1. Introduction

Fibre-reinforced composite materials are used extensively in the aerospace industry because of their light weight, superior corrosion resistance and improved fatigue properties when compared to metals. However, the manufacturing costs, production rates and damage resistance are current challenges faced by the composite industry. Three-dimensional (3D) woven composites have better through-the-thickness properties in comparison to two-dimensional (2D) laminates; they show damage resistance, high inter-laminar fracture toughness and reduced notch sensitivity that demonstrate a better damage tolerance. 3D fabrics were introduced to produce structural composites capable of withstanding multidirectional stresses.

Monitoring of acoustic emission (AE) during mechanical loading is an effective and widely used tool in the study of damage processes in glass fiber-reinforced composites. This study provides further insight into the AE monitoring of 3D AI glass fibre composites. Tests were performed with piezoelectric sensors bonded on a tensile specimen acting as passive receivers of AE signals. These signals are carefully analysed to identify resin cracks in the warp yarn and relate to crack density.

1.1. Damaged Monitored by Acoustic Emission in Composite Materials

AE is a passive SHM technique that can be used for many applications. When crack grows, energy is released at the crack tip in form of waves. AE sensors can be used to measure these waves. Several sensors in combination can be used to estimate the severity of the crack and its location. Most publications show results from fatigue cracks in bulk materials and qualitative results from real structures. However, there is limited literature presenting quantitative results from plate-like structures and a lot of the experiments are based on simulated AE sources, e.g., pencil lead breaks [1]. One aim of this paper is to analyse the elastic waves generated from transverse cracks (TC) in a 3D angle interlock composite structures subjected to tensile loading. FEM can be used to model the AE waves from transverse crack and it can provide a better understanding of the AE generated from TC in composite plates.

The AE method allows the detection and location of damage using specific localisation algorithms. Knowledge of the propagation velocity and attenuation of the AE wave is required. However, contrary to metallic material, the anisotropic nature of composite material gives a large range of propagation velocity due to fibre orientation. Moreover, the attenuation of the AE waves is more complex than in a homogeneous material [2]. In addition, in a same composite material, wave attenuation is more significant in cracked than in healthy state, which will complicate the signal processing after few damage modes have developed, especially for the amplitude distribution. Qualifying damage started first in 2D composites and Mehan and Mullin in 1968 [3] managed to identify three basic failure mechanisms: (i) fiber fracture; (ii) matrix cracking; (iii) and fibre/matrix interfacial debonding. The authors reported the application of AE in composites in 1971 [4], discriminating audible types for these three basic damage modes using an AE system. After forty years, Godin et al. [5] conducted mapping of cross-ply glass/epoxy composites during tensile tests. They have classified four different acoustic signatures of failure and determined four conventional analyses of AE signals.

Typical waveforms with A-Type (slow increase times at about 10–20 μs) signals associated with matrix cracking, B-Type (sharp rising, lasted for 10 μs and abruptly decreasing) with fibre/matrix interface de-bonding, C-Type associated with fibre failure, and D-Type (long rising times, high amplitudes, and very long durations) with delamination [5]. The most popular methods to identify damage are identification by signal amplitude distribution (signal strength) and by signal frequency.Table 1 and Table 2 show a comparison between the amplitude and the frequency distribution model that were encountered in the literature.

Table 1. Amplitude distribution according to the damage mechanism in composite materials.

Ref.	Matrix Cracking	Interface Decohesion (Fibre/Matrix)	Fibre/Matrix Friction and Fibres Pull-Out	Fibres Breakage
[6]	30–45 dB	45–55 dB	–	>55 dB
[7]	60–80 dB	70–90 dB	–	–
[8]	50 dB	–	–	–
[9]	40–70 dB	–	–	60–100 dB
[10]	40–55 dB	–	>80 dB	–
[11]	33–45 dB	50–68 dB	69–86 dB	87–100 dB
[12]	40–78 dB	72–100 dB	–	95–100 dB
[13]	40–55 dB	60–65 dB	65–85 dB	85–95 dB
[5]	35–80 dB	50–80 dB	70–100 dB	–
[14]	<70 dB	<60 dB	–	–
[15]	35–55 dB	55–100 dB	–	35–80 dB
[16]	40–60 dB	50–70 dB	80–100 dB	80–100 dB

Table 2. Frequency distribution according to the damage mechanisms in composite materials.

Ref.	Matrix Cracking	Interface Decohesion (Fibre/Matrix)	Fibre/Matrix Friction and Fibres Pull-Out	Fibres Breakage
[17]	50–150 kHz	–	–	140–180 kHz
[18]	30–150 kHz	30–100 kHz	180–290 kHz	300–400 kHz
[19]	80–130 kHz	–	250–410 kHz	250–410 kHz
[14]	~300 kHz	–	300 kHz	>500 kHz
[20]	50–180 kHz	220–300 kHz	180–220 kHz	>300 kHz
[21]	90–110 kHz	–	200–300 kHz	>420 kHz
[22]	<50 kHz	200–300 kHz	500–600 kHz	400–500 kHz
[23]	~140 kHz	~300 kHz	–	~405 kHz
[24]	200–600 kHz	200–350 kHz	0.7–1.1 MHz	>1.5 MHz
[15]	50–80 kHz	50–150 kHz	–	150–500 kHz

All of these studies show the difficulty of identifying damage modes for 2D composites and becomes more complicated for 3D woven composites. Only a small amount of investigation has been reported for monitoring evolution of damage and ultimate failure in 3D woven composites. Li et al. [15] studied AE signals for 3D non-crimp orthogonal woven glass/epoxy composites from cluster analysis point of view. These clusters are based on different parameters of peak amplitude, peak frequency, and RA value (rise time divided by peak amplitude). From their investigation, cluster 1 (low frequency, low amplitude events) and 2 (moderate frequency, low amplitude) is correlated to matrix cracking, cluster 3 (low to moderate frequency with high amplitude) with fibre and matrix de-bonding, and cluster 4 (high frequency) with delamination and fibre breakage. Lomov et al. [25] investigated AE response in 3D non-crimp orthogonal woven carbon/epoxy composites undergone damage.

However, identifying cracking in the matrix or fibre in addition to delamination need to be investigated further if AE is to be used as an inspection tool in SHM of 3D woven composites. Hence, the present study (qualitative and quantitative) of 3D angle-interlock woven composite damages using AE piezoelectric sensors is undertaken. As these structural woven fabrics are attracting the attention of the aerospace industry, the monitoring of initiation and progression of transverse matrix cracking is of considerable interest and importance, since they can lead to delamination and fibre breakage, which result to ultimate failure.

1.2. Guided Waves

Guided waves are very widespread in SHM applications: Guided waves are important for SHM applications because they have the ability to travel without much energy loss over large areas. This property makes them well suited for ultrasonic inspection of bridges, aircraft, ships, missiles, pressure vessels, pipelines, etc. In plates, ultrasonic guided waves propagate as Lamb waves and as shear horizontal waves (SH). Ultrasonic guided waves in plates were first described by Lamb (1917). A detailed study of Lamb waves has been given by Viktorov [26], Achenbach [27], Graff [28], Rose [29] and Dieulesaint and Royer [30]. Lamb waves are of two varieties, symmetric modes (S0, S1, S2...) and anti-symmetric modes (A0, A1, A2...). At low values of the frequency-thickness product, *fd*, the first symmetric mode, S0, resembles axial waves whereas the first anti-symmetric mode, A0, resembles flexural waves. The choice of Lamb waves is justified by their many advantages; they have the power to energize the entire thickness of the plate and offer the possibility of detecting internal defects at various depths. However, Lamb waves present some difficulties: they are dispersive, and also several modes can propagate at different speeds at a given frequency. Work has been done to establish analytically the dispersion curves in isotropic plates [30,31], to validate the results experimentally and to study the effect of dispersion over long distances [32]. Lamb wave propagation was used by many authors [33,34,35] using piezoelectric disks as transmitters and receivers to measure the changes in the signal received from a structure having a defect. However the signal processing is complex due to multiple reflections. Today the majority of work concerns the propagation of Lamb waves in thin isotropic structures. For this reason, it is very important to study the Lamb wave propagation from an acoustic emission point of view in 3D composite materials to understand the difficulties in analysing these waves in order to be able to qualify and quantify the defects in such structural configurations.

2. Materials Presentations and Experimental Set-Up

In this study, a 3D angle interlock (AI) S2 glass woven composite plate with through-thickness binding was infused using bi-functional epoxy resin (LY564) and hardener (XB3486) supplied by Huntsman. In the AI configuration, the binder goes all the way through-the-thickness and then returns back. According to the binding pattern, shown in Figure 1 one binder yarn is inserted after every three layers of weft (yarn). This structure consists of 4 layers of warp (fibres parallel to weaving direction or at 0°) and 3 layers of weft (fibres transverse to weaving direction or at 90°), which are held together by the binders (through-thickness fibres) inserted in the weft direction at regular intervals as illustrated inFigure 1.

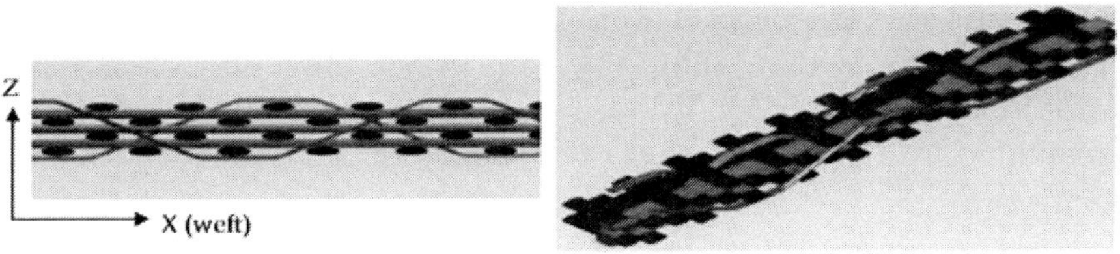

Figure 1. A schematic of 3D Angle Interlock Woven Composite (through thickness and planar view) (orange: weft; black: warp; green: binder yarn) (Binder yarn goes all the way through-the-thickness, z-axis, and then returns back).

Tensile testing was carried out according to ASTM standard D3039 [36], on specimens 250 mm long (with a gauge length of 50 mm) and 25 mm wide. The tensile load was applied in the weft direction. A non-contact video extensometer was used to measure the strain developed while the specimen was loaded in an Instron 5982 R2680 testing machine (Norwood, MA, USA). Three piezoelectric wafer active sensors (PWAS) bonded on the specimen were acting as AE receivers, Figure 2.

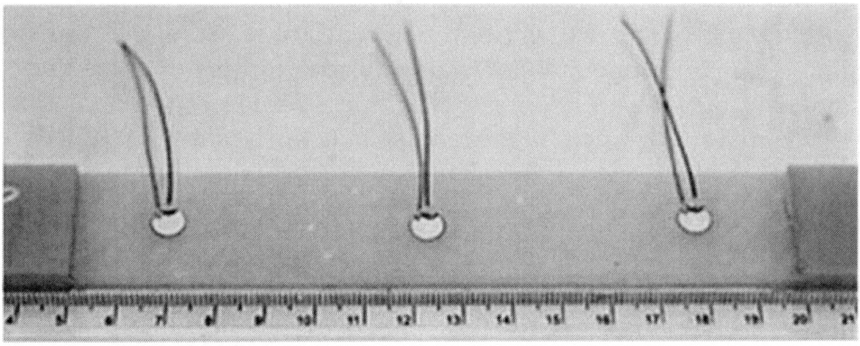

Figure 2. PWAS bonded on a 3D angle interlock glass fibre tensile specimen for acoustic emission.

To develop only transverse cracks, the specimen was loaded up to 20% of its ultimate strength (σ_f). During loading, acoustic emission signals were recorded and the PWAS were able to pick up AE signal of good strength at a frequency range 100–700 kHz. The acquisition of the signals was performed using software 'AEWin' from Mistras with a sampling rate of 10 MHz and 20 dB pre-amplification. The AE PWAS sensors used in this study were provided by Steminc, further details in [37].

3. Angle Interlock Cracking Simulation

Fibre reinforced composite materials exhibit mostly a linear elastic behaviour similar to brittle materials up to the final failure specially when loaded along the fibre direction in tension. This mainly occurs because the most significant contribution for the load carrying capacity of these materials depends on the longitudinal fiber properties and strength. Even if some progressive failure occurs in the matrix or transverse cracking, still composites can carry the load up to the fiber failure along the loading direction. From this perspective, linear elastic fracture mechanics can be employed to describe and analyse the fracture "cracking" of fiber reinforced composites [38]. Any finite fracture that occurs in a composite material is governed by the first law of thermodynamics. The energy dissipated due to crack formation normalized by the surface area of the newly formed crack is known as the energy release rate (G). Transverse cracking and local delamination are two common types of cracking mechanisms that occur in composite materials. In order for any of these matrix cracking mechanisms to exist [38], the strain energy release rate associated with each damage mechanism (G) should exceed its critical strain energy density "toughness" (Gc). Therefore, the question always is how to determine the energy release rate (G) for heterogeneous materials like composites. The strain energy release rate for composite materials is calculated as [39]:

$$G = -\frac{\Delta E}{\Delta A} \tag{1}$$

where ΔE is the strain energy released due to the cracking formation. This is determined by subtracting the strain energy density of a cracked cell from the strain energy density of non-cracked cell while ΔA represents the area of the cracked surface. Strain energy release rate actually defines the potential locations for crack formation along the yarn or its cross section. Cracks are more likely to form in locations where the strain energy release rate is high.

For composite materials, the strain energy density can be calculated as function of the applied strain/ stress. Therefore, the strain energy density components can be calculated as follows [40]:

$$e_{ij} = \frac{1}{2V} \int_{v} \frac{\sigma_{ij}^2}{E_{ij}} \tag{2}$$

where V is the volume of the (ply/yarn/laminate) determined as the cross-sectional area multiplied by the thickness, σij is the ij component of stress and Eij is the corresponding Young's modulus ($i = j$) or Shear modulus ($i \neq j$).

Figure 3 is a graph to illustrate the theory behind the finite fracture mechanics. The toughness of the material for a specific cracking mechanism (Gc) is a material property which is constant while the energy release rate increases with increasing applied stress/strain. Once the energy release rate associated with a specific cracking mechanism exceeds the critical value, crack formation and damage evolution starts.

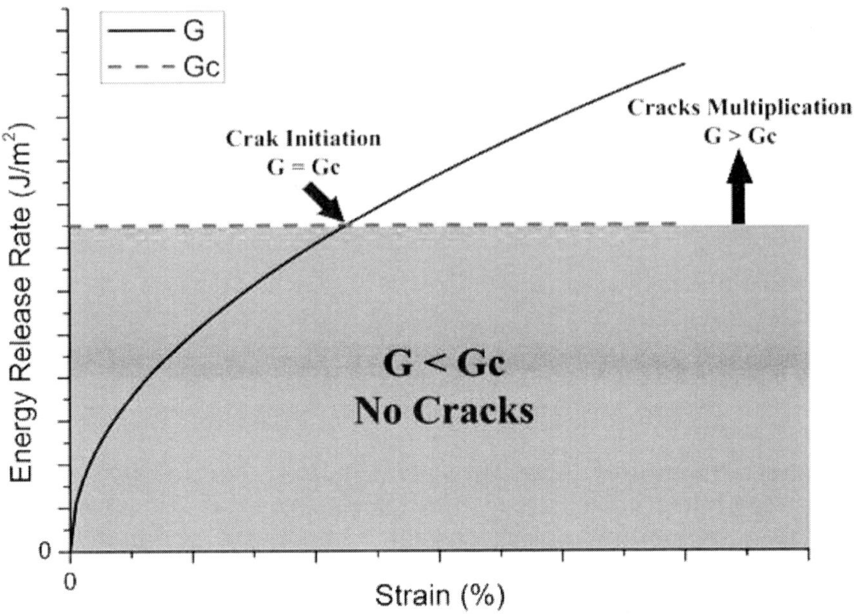

Figure 3. Graphical representation of the finite fracture mechanics theory.

On more issue regarding the fracture of composite materials is that the fracture occurs due to multiplication of cracking events rather than growth of a single crack. Therefore, the fracture response of composite materials is more like discrete instantaneous crack propagation. For further details about the application of finite fracture mechanics of composite materials, the reader is referred to [38].

The 3D Angle Interlock Woven Composite (3DAWC) (Figure 1) is modelled as a (0/90) cross-ply laminate since the crimp mostly occurs at the interlacement points between the weft and binder yarns [40]. In order to check the effect of this simplification on the in-plane properties of the 3DAWC, analytical homogenization technique "orientation averaging model" is used to calculate approximately the elastic material properties [41,42] and compare it with the measured data obtained. As shown inTable 3, good agreement between the experimental and analytical model is obtained while the last column represents the difference between the calculated values with and without the binder yarns, confirming that the z-yarns have negligible effect on axial stiffness. This result justifies the representation of the 3D woven architecture by a cross-ply (0/90) laminate used in the AE simulation, see Section 4.

Table 3. Elastic material properties of 3D AI woven composites.

Mechanical Data	Experiment	With Binder	Without Binder	Difference (%)
E_1	18.52 ± 0.87	17.85	17.33	2.91
E_2	24.83 ± 1.51	24.00	23.48	2.16
E_3	–	12.74	11.00	13.65
G_{12}	–	5.18	4.95	4.50
ν_{12}	–	0.31	0.32	0.68
V_F (%)	$V_F = 50.35 \pm 0.26$; $V_F(\text{warp}) = 31.21 \pm 0.26$; $V_F(\text{weft}) = 15.38 \pm 0.36$; $V_F(\text{binder}) = 3.05 \pm 0.33$			

A larger impact of the through-the-thickness reinforcement is expected on the interlaminar fracture toughness rather than in-plane stiffness properties. An almost 14% increase in E33 modulus is predicted when the binder yarns are considered in the analysis.

To determine which constituent part of the 3D woven will experience cracking in the case of uniaxial tension, strain energy density components are calculated for the 3D AI woven composites unit cell when applying 1% strain along the weft direction. The finite element model is run using the COMSOL Multiphysics software package. Figure 4 shows that the transverse component eTT of the strain energy density is the highest when compared to the longitudinal eLL and shear eLT components. This implies that the strain energy release rate for the transverse component is the one that leads to matrix cracking in the weft yarn under this loading condition. In addition, having a constant energy release rate along the whole yarn length, it suggests that there is no preferable location within the yarn for the crack to start from. This also means that once a crack is initiated in the yarn, it grows instantaneously through the thickness and along the whole yarn length. The complete study of damage mechanisms is well explained and characterised in references [43,44].

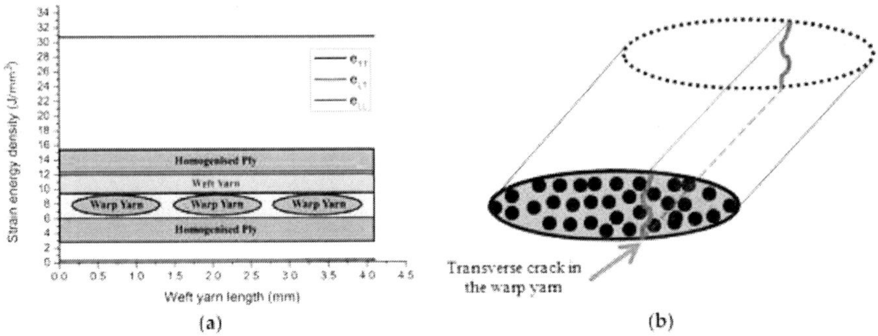

Figure 4. (**a**) Strain energy release rate along weft yarn (TT: Transverse component; LT: shear component; LL: axial component); (**b**) crack on a warp yarn cross section (Transverse crack).

Matrix cracking is a phenomenon that generates a motion which is essentially in plane. The motion of the crack faces is parallel to the plane of the specimen. It can thus be expected that matrix cracks will generate AE waves which contain a predominant extensional mode. Fibre fracture follows the same general behaviour and should therefore also be characterised by a large extensional mode [45].

A delamination is a damage phenomenon that generates a motion, which is essentially out of plane. In this case, the motion is perpendicular to the plane of the plate. Delaminations should thus generate AE waves, which contain a dominant flexural mode. Fibre/matrix debonding follows the same behaviour and should also be characterised by a large flexural mode. It should be noted that delamination and fibre/matrix debonding can be also driven by shear stresses where there is no crack opening but crack sliding making it more difficult to detect non-destructively.

4. Acoustic Emission Simulation

Simulation of AE was realised using the ABAQUS/implicit software (Dassault Systèmes, Vélizy-Villacoublay, France), which has multi-physics piezoelectric elements. FEM modelling was used to simulate the elastic wave emitted by the transverse crack growth. These can be used to compare with the results obtained from the experiment. The ABAQUS model is shown in Figure 5. This structure, consisting of 4 layers of warp (at 0°), 3 layers of weft (or at 90°), and held together by the binders (through-thickness fibres) are homogenised. Two elements per ply are used. Eight nodes linear piezoelectric brick element were used to simulate the PWAS. Implicit solver methods of solution are used in order to simulate the real voltage/amplitude received signal [46]. The use of multi-physics finite element method (MP-FEM) is explored to model the reception of the elastic wave as electric signal recorded at a PWAS receiver (R-PWAS).

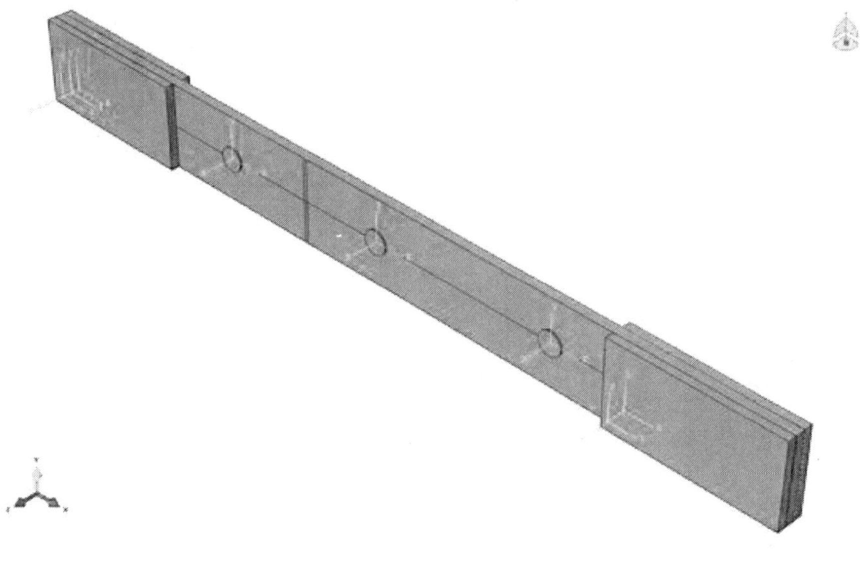

Figure 5. ABAQUS model of the homogenised 3D woven composite with 3 PWAS bonded on the top to record the AE events from the surface simulated transverse crack.

The piezoelectric material properties were assigned to the PWAS as described in reference [37]:

$$[C] = \begin{pmatrix} 97 & 49 & 49 & 0 & 0 & 0 \\ 49 & 97 & 44 & 0 & 0 & 0 \\ 49 & 49 & 84 & 0 & 0 & 0 \\ 0 & 0 & 0 & 24 & 0 & 0 \\ 0 & 0 & 0 & 0 & 22 & 0 \\ 0 & 0 & 0 & 0 & 0 & 22 \end{pmatrix} (\mathrm{GPa}) \quad (3)$$

$$[\varepsilon] = \begin{pmatrix} 947 & 0 & 0 \\ 0 & 947 & 0 \\ 0 & 0 & 605 \end{pmatrix} \times 10^{-8}\,(\mathrm{F/m}) \quad (4)$$

$$[e] = \begin{pmatrix} 0 & 0 & 0 & 0 & 12.84 & 0 \\ 0 & 0 & 0 & 12.84 & 0 & 0 \\ -8.02 & -8.02 & 18.31 & 0 & 0 & 0 \end{pmatrix} \left(\mathrm{C/m^2}\right) \quad (5)$$

where [C] is the stiffness matrix; [ε] is the dielectric matrix and [e] is the piezoelectric matrix. PWAS has a density of $\rho = 7600$ Kg/m^3, diameter of 7 mm, and thickness of 500 μm. The 3D composite properties are shown in Table 3 and the Rayleigh damping coefficients from reference [2] are used. It should be noted that these Rayleigh damping coefficients may have an effect on the wave amplitude of the signal but not the shape of the waveform, which is used in characterizing the damage mode.

The maximum frequency of interest was chosen at around 600 KHz. For 600 KHz, a time interval of 0.1 μs and an element size about 0.5 mm in the composite plate are required to achieve an error on wave velocity below 5% [46,47]. A step excitation was used as shown in Figure 6a. To simulate the energy released by the transverse crack a two-point source force was applied between PWAS#1 and PWAS#2 at the surface of the specimen as illustrated in Figure 6b. A shear force, parallel to the crack could also be used, but would have no effect on the shape of the signals received by the PWAS. The end of the specimen is fixed to represent the real boundary conditions of the tensile test. However, the tensile load is not simulated.

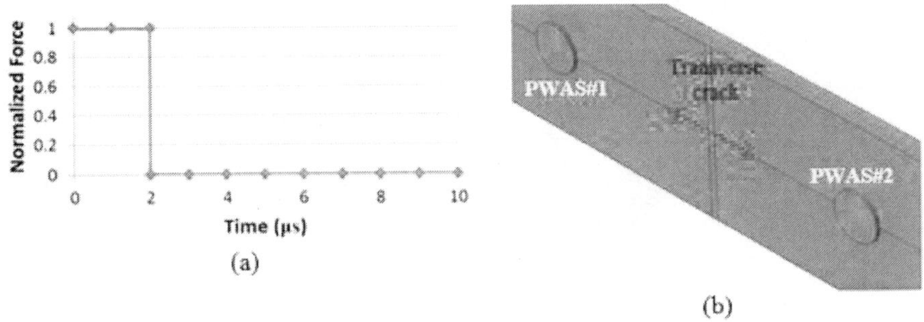

Figure 6. (**a**) Source function used: at time zero the force step up from 0 to a nominal value 1, and then return to 0 at 2 μs; (**b**) two-point source force to simulate the energy release by the transverse crack.

5. Results and Discussion

5.1. Multi-Physics Finite Element Simulation

Figure 7 shows image snapshots of overall displacement amplitude of the guided wave pattern in the plate taken at 10-μs intervals. Multiple guided waves modes are present. At *t* = 10 μs, one sees the waves just starting from the transverse crack. By *t* = 20 μs, most of the wave has already being reflected from the edges of the tensile specimen which will complicated the analysis of the received signal due to Lamb waves mode conversion.

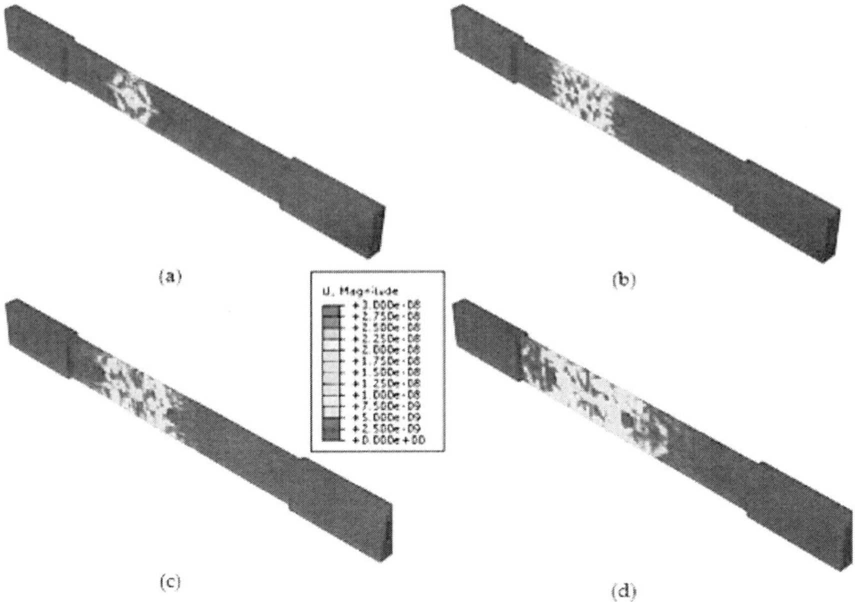

Figure 7. Snapshot of the MP-FEM simulation of guided waves generate by a pair of point forces simulating an acoustic emission by the transverse crack in a 3D angle interlock composite tensile specimen at (**a**) 10 μs; (**b**) 20 μs; (**c**) 30 μs; (**d**) 40 μs.

The simulated AE signal caused by the simulated transverse crack excitation as captured at PWAS#1, 2, and 3 is shown in Figure 8. The magnitude of the received signal from PWAS#3 (in green) decreased dramatically due the damping effect introduced in the model.

To better understand these signals, the discrete wavelet transform (DWT) is used. The DWT of a time signal s(t) is the result of the convolution product between the signal s(t) and a family of “daughter wavelets” γm.k(t),

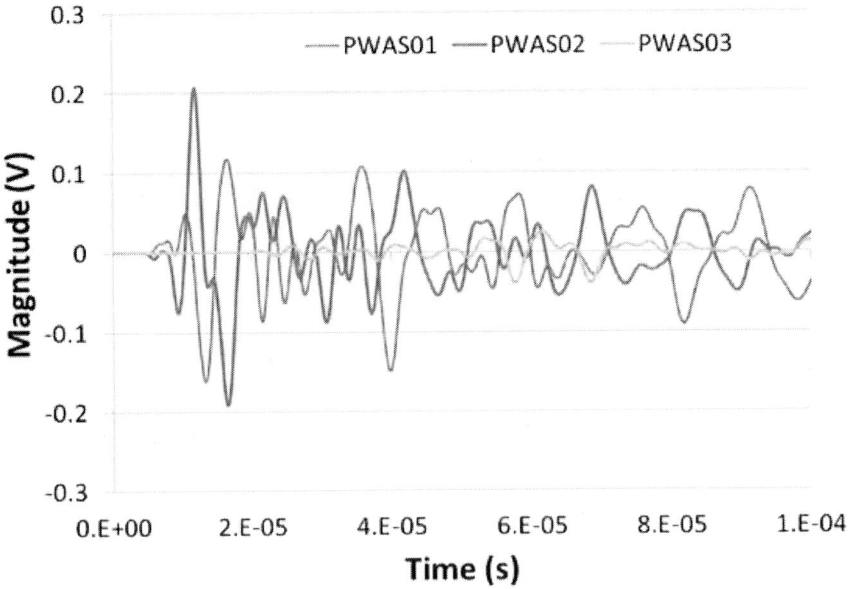

Figure 8. Simulated signal received: Output voltage against time for PWAS#01, 02, and 03.

$$DWT_{m,k} = \int_{0}^{\infty} s(t)\gamma_{m,k}(t)dt \tag{6}$$

The main particularity of the DWT is that the result obtained with each daughter wavelet corresponds to the time behaviour of the signal in a frequency band corresponding to dilatation factor *m*. Each response is called the decomposition level. A number of different bases have been proposed to construct a family of wavelets. A good solution for analysis and decomposition can be obtained with the Morlet wavelet. The application of discrete wavelet analysis to the acquired AE signals resulted in its decomposition into six different levels. Each level represents a specific frequency range, and the frequency range increases with increasing wavelet level. The decomposed AE signals in level 1 to 5 are shown in Figure 9 for the PWAS#01.

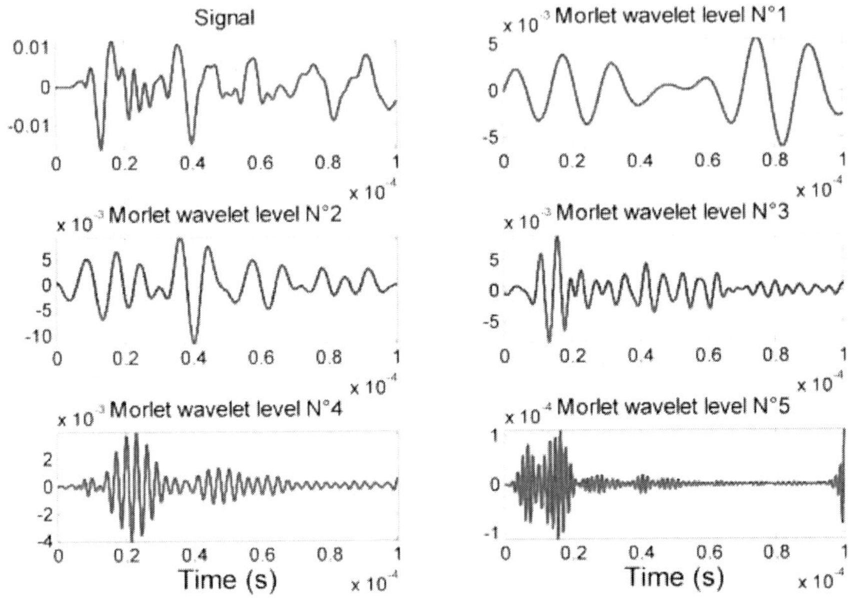

Figure 9. Discrete wavelet transform of the simulated signal received by the PWAS#1.

The Fourier spectrum of the Figure 9 signals is shown in Figure 10. The frequency spectra for DWT levels 1 through 5 are centered at about 68 kHz, 120 kHz, 200 kHz, 340 kHz, and 650 kHz, respectively. At frequencies 68 kHz, 120 kHz, and 200 kHz (Morlet wavelet levels 1 and 2), three modes exist, the fundamental symmetric mode (S0), the fundamental anti-symmetric mode (A0), and the fundamental shear mode (SH0). However, with the PWAS receiver geometry and properties, the SH mode cannot be caught by these sensors [2]. Moreover, based on the tuning study, at 68 kHz the amplitude of the A0 mode is much higher than the S0 mode, and its travel speed is slower. At 120 kHz, the amplitude of A0 and S0 are almost the same, and at 200 kHz, the amplitude of the S0 is higher than the A0. To conclude, the component at low frequency (below 140 kHz) is dominated by the fundamental antisymmetric mode A0. At 340 kHz (Morlet wavelet level 3), four modes are existent, S0, A0, A1 and S1; at 650 kHz (Morlet wavelet level 4), six modes are present, S0, S1, S2, A0, A1, and A2. Therefore, at these frequencies, the distinction of the different wave packets and the signal processing are very complex. Moreover, the amplitude is distributed such that it is the highest in level 1 and lowest in level 5 as shown in Figure 9. The FFT of the original signal shows that the amplitude of the signal is higher for the frequency lower than 160 kHz, which mean that the transverse crack develops more flexural (i.e., A0) than extensional (i.e., S0) motion.

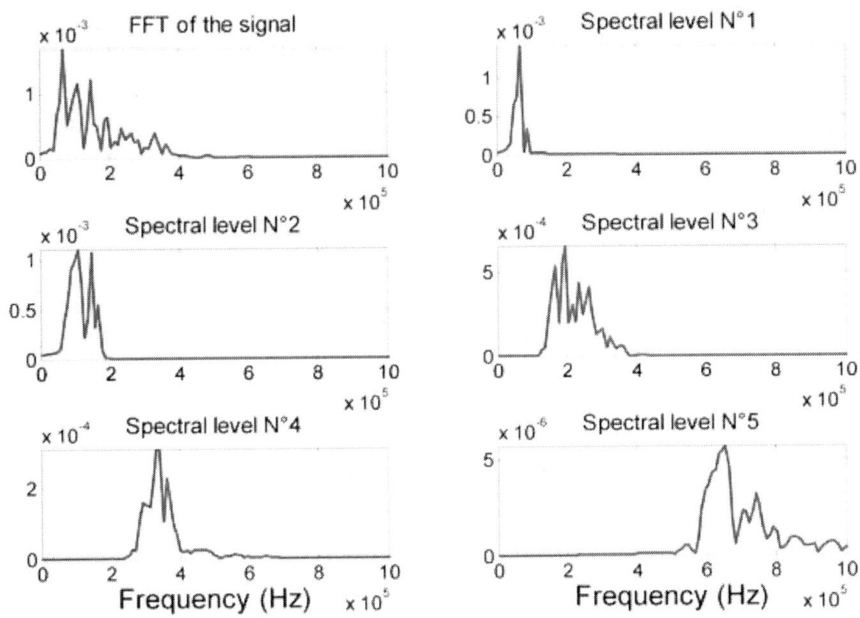

Figure 10. Frequency spectra for the different wavelet level (PWAS#1).

However, Surgeon and Wevers [42] mentioned that matrix cracks will generate AE waves which contain a predominant extensional mode (i.e., S0 mode). It might be explained by the symmetry of the transverse crack, which is maybe not the case in our experiments.

Figure 11 shows the continuous wavelet transform (CWT) magnitude as a function of frequency versus time. The CWT were calculated with AGU-Vallen Wavelet, a freeware software program [48]. This program has a Gabor function as the "mother" wavelet. Figure 11 shows the analytical dispersion curves with the three lowest modes (S0, A0, and A1) superimposed on the CWT plot. The colour scale is a linear scale with black representating the highest magnitude and white the lowest or zero-magnitude region. Clearly, Figure 11 shows the presence of AE signal energy in portions of mainly two modes, A0 and S0. The CWT shows how the signal energy is distributed as a function of frequency, time (or group velocity), and mode. Figure 11 shows that the simulated AE source has the greatest concentration (most black color) of energy is the fundamental anti-symmetric mode A0 in a frequency range of 50 to 250 kHz. Another large amplitude region of the CWT is the part of the fundamental symetric mode S0 in a frequency range 50 to 300 kHz. This demonstrates that the AE signal energy is not uniformly distributed between the modes; it is also not uniformly distributed as a function of frequency along each of the dominant modes.

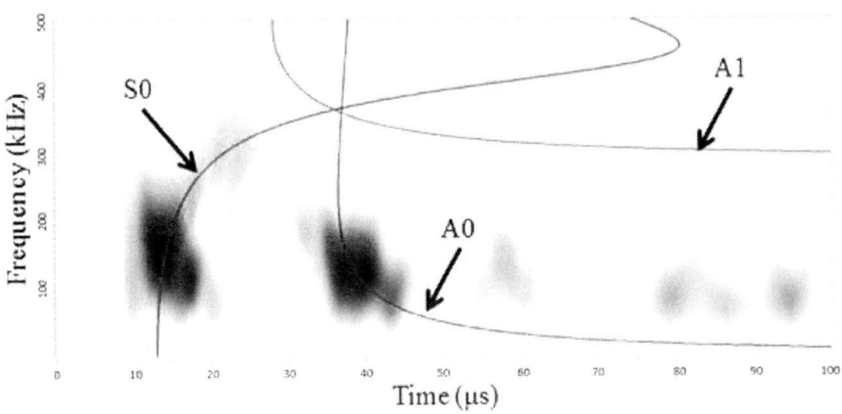

Figure 11. Superimposed symmetric mode and anti-symmetric modes after converting group velocity to time based on the propagation distance. Light and dark grey correspond to simulated AE activity.

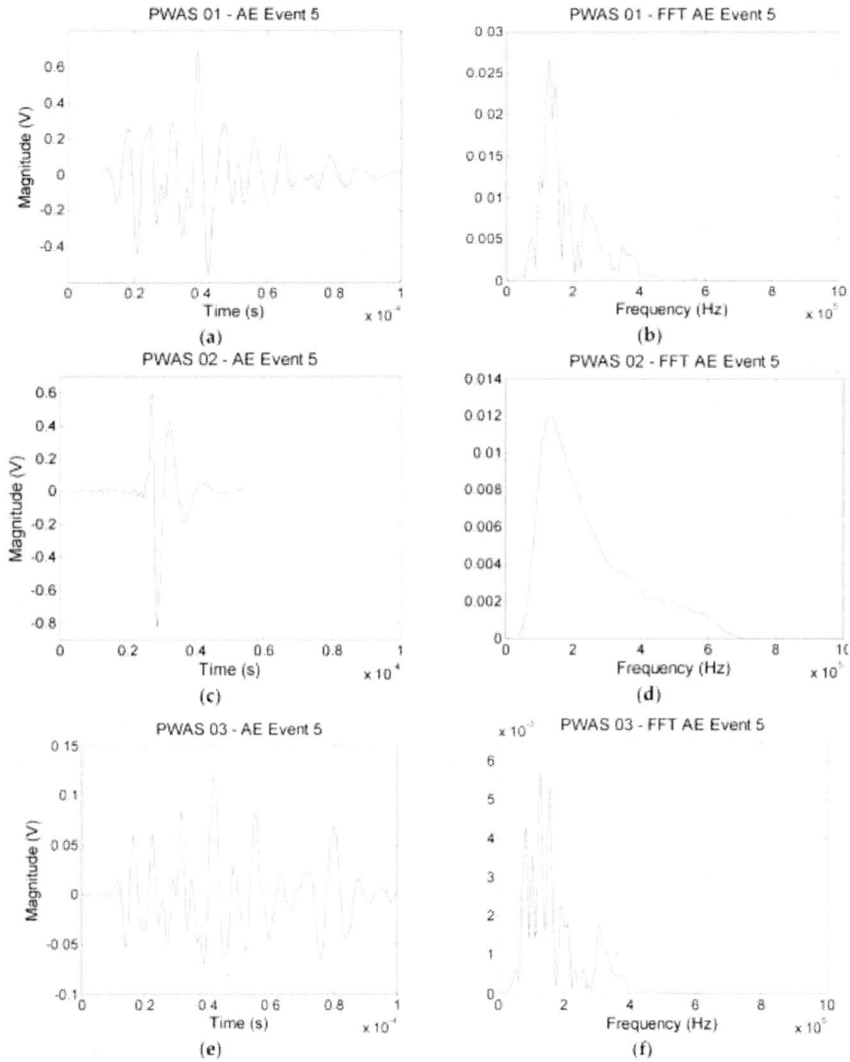

Figure 12. Typical experimental AE waveforms and Fourier Transform from a transverse crack in 3D AI recorded from (**a**,**b**) PWAS#1; (**c**,**d**) PWAS#2; (**e**,**f**) PWAS#3.

The above discussion proves that the waveforms features (duration time, amplitude, time-frequency spectrum) are useful to illustrate the characteristics of AE signal and distinguish the different AE signals associated with various possible failure modes in the specimens. Moreover, PWAS#2 and PWAS#3 obtained similar trend to the PWAS#1.

5.2. Experiments

As mentioned in Section 3, at this applied tensile load only transverse cracking occurs in the studied specimen. Figure 12 shows typical AE waveforms received by the PWAS#1, #2, and #3, and the associated Fourier transform.

In this particular example, the transverse crack occurs closer to PWAS#2 than the other sensors. This signal looks sharper and stronger than those obtained by PWAS#1 and #3. Masmoudi et al. [12] classified these very energetic signals with amplitude above 94 dB to fibre breaking. However, in theory, no fibre breakage should occur, only transverse crack in the warp yarn should develop as previously simulated. In the next section, the stress amplification factor (SAF) is introduced to explain this typical fibre breakage waveform. The amplitudes of this particular event are 96, 98, 81 dB for PWAS#1, #2, and #3, respectively. The amplitude decreases with the travel length due to the high damping coefficient in this 3D AI composite materials.

Figure 13 shows the CWT magnitude as a function of frequency versus time and shows the anlaytical dispersion curve with the three lowest modes (S0, A0, and A1) superimposed on the CWT plot of the typical AE waveforms recorded from PWAS#1, #2 and #3. The colour scale is a linear scale with black representating the highest magnitude and white the lowest or zero-magnitude region. The CWT shows how the signal energy is distributed as a function of frequency, time (or group velocity), and mode.Figure 13a shows the presence of AE signal energy in portions of mainly two modes, A0 and S0 for the PWAS#1 which is in agreement with our MP-FEM results shown in Figure 11. The experimental AE source has the greatest concentration of energy is the fundamental flexural mode A0 in a frequency range of 80 to 300 kHz (the simulated AE event is in a frequency range of 50 to 200 kHz for the A0 mode). Another large amplitude region of the CWT is the part of the fundamental extensional mode S0 in a frequency range 110 to 220

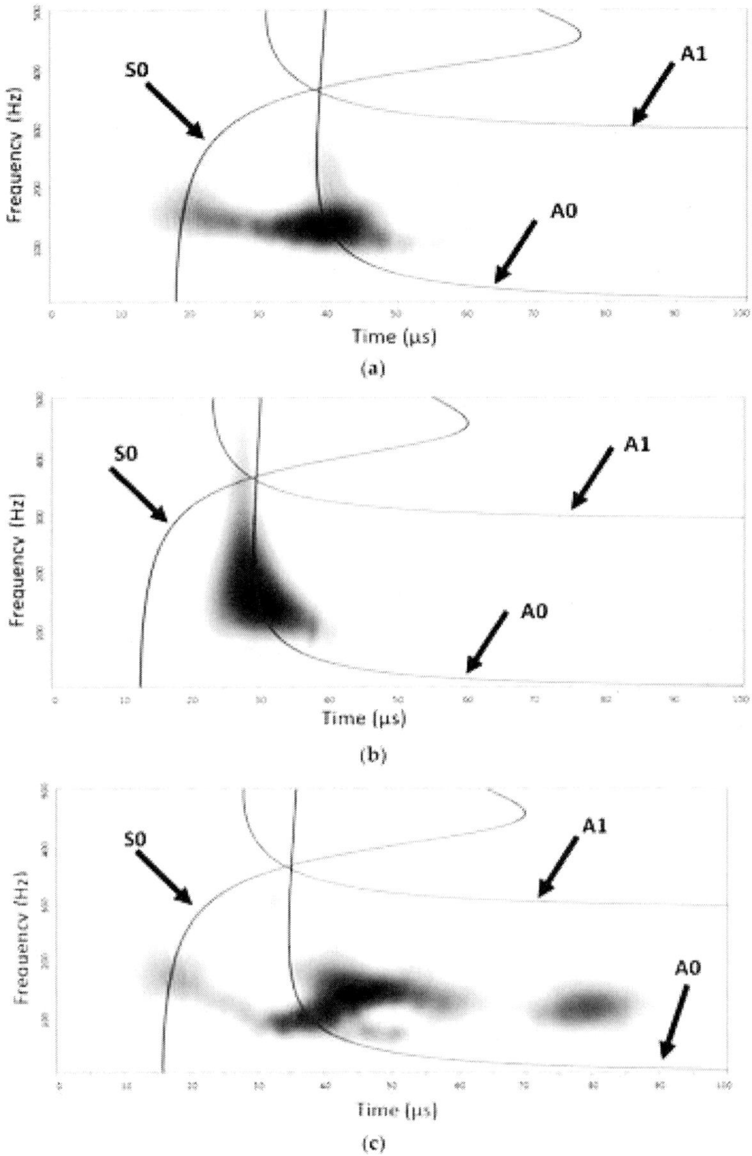

Figure 13. Superimposed symmetric mode and anti-symmetric modes after converting group velocity to time based on the propagation distance for the experimental received signal: (**a**) PWAS#1; (**b**) PWAS#2; (**c**) PWAS#3.

kHz (the simulated AE event is in a frequency range of 50 to 300 kHz for the S0 mode). Figure 13b shows the presence of AE signal energy in portions of only one mode, A0 for the PWAS#2. This experimental AE source is the fundamental flexural mode A0 in a frequency range of 80 to 500 kHz with a higher concentration between 120 to 250 kHz. During this typical event, damage occurs close to PWAS#2 and so the wave does not have time to travel over long distance. Moreover, this waveform is assimilited to a micro-fibril breakage (binder yarn) with very high energy which shadow all the reflection waves from the edge. Figure 13c shows the presence of experimental AE signal energy in portions of mainly two modes, A0 and S0 for the PWAS#3.

Figure 13c shows that the AE source has the greatest concentration of energy is the fundamental flexural mode A0 in a frequency range of 60 to 230 kHz (the simulated AE event is in a frequency range of 50 to 200 kHz for the A0 mode). Another large amplitude region of the CWT is the part of the fundamental extensional mode S0 in a frequency range 130 to 250 kHz (the simulated AE event is in a frequency range of 50 to 300 kHz for the S0 mode). Because the experimental AE event occur far away from the PWAS#3 several reflections are also visible. This demonstrates that the AE signal energy is not uniformly distributed between the modes; it is also not uniformly distributed as a function of frequency along each of the dominant modes.

In summary, it seems that transverse crack (simulated and experimental) generates a predominant flexural mode A0 and also a less energetic extensional mode S0. Moreover, the micro-fibril breakage (in the binder yarn) at the tip of the transverse crack (typical waveform—Figure 12c) generates only the fundamental flexural mode A0. This conclusion is in disagreement with previous study [45]. It might be explained by the non-symmetry of the damage which is maybe not the case in the others experiments.

Moreover, the frequency of these signals show clearly two major components, the first one between 70 and 180 kHz and the second one between 200 and 400 kHz for PWAS#1 and #3.

The high frequency and the low frequency component correspond to the wave's extensional mode S0 and to the flexural mode A0, respectively, as showed in the MP-FEM simulation. This flexural mode A0 has higher amplitude than the extensional S0 mode. It seems that the transverse cracks generate more flexural motion than extensional motion. This presence of a flexural mode would indicate that the crack does not develop symmetrically about the mid-plane of the 3D AI laminate. The crack initiation for the loading in weft direction occurs in the range of applied strain 0.07%...0.1% (Figure 14, showing the data for weft direction of loading), a relatively low level of strain. The amplitude for each AE event (i.e. transverse crack) is between 60 and 100 dB. The signals with lower amplitude were assimilated into noise.

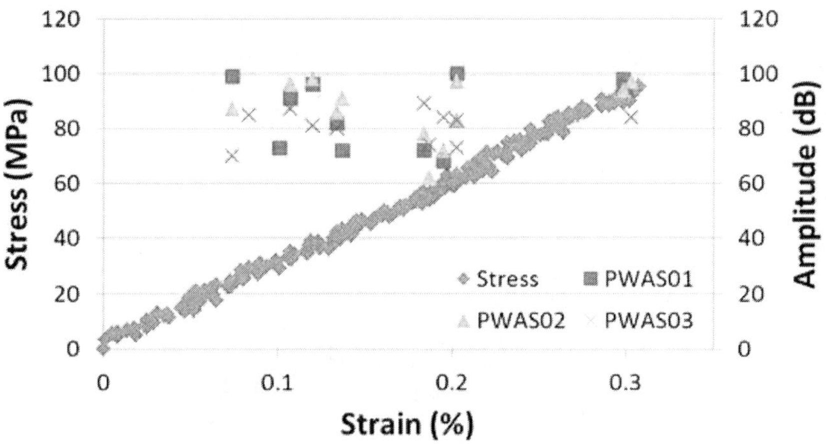

Figure 14. Applied stress-strain curve and the PWAS amplitude for each AE events (transverse cracks and micro-fibril breakage). Ultimate failure strain = 1.3%.

These experimental and simulated results have proven that transverse matrix cracking signals do exhibit a clear fundamental flexural A0 mode. In most cases, however, the extensional mode was also clearly present. For the transverse matrix crack signals this is caused by their asymmetric growth through the thickness. Matrix cracks most often initiate at one of the outer plies and grow through the thickness to the other side of the specimen. These results in a particle motion which is in plane, but asymmetric about the mid-plane, thus resulting in a flexural mode. The large flexural mode observed during this test can be explained by the same principle: transverse cracks will occur preferably in the zone of maximum tensile stress. AE waves generated there will thus cause an in plane motion, but the motion will be asymmetric about the mid-plane. This will again result in a flexural component.

5.3. Stress Amplification Factor

On the micro-mechanical analysis, the external applied stress and the local stress within the material is not the same due to the difference in the material properties of the material constituents. A random fibre distribution in a yarn can be simplified by a unit cell of a hexagonal array distribution. When this unit cell is subjected to an external load as shown in Figure 15, the fibre and matrix will experience different stresses resulting in a stress concentration within the unit cell. Therefore, it is obvious that if an external uniform unit load is applied on the boundary, the stresses within the unit cell are not unity.

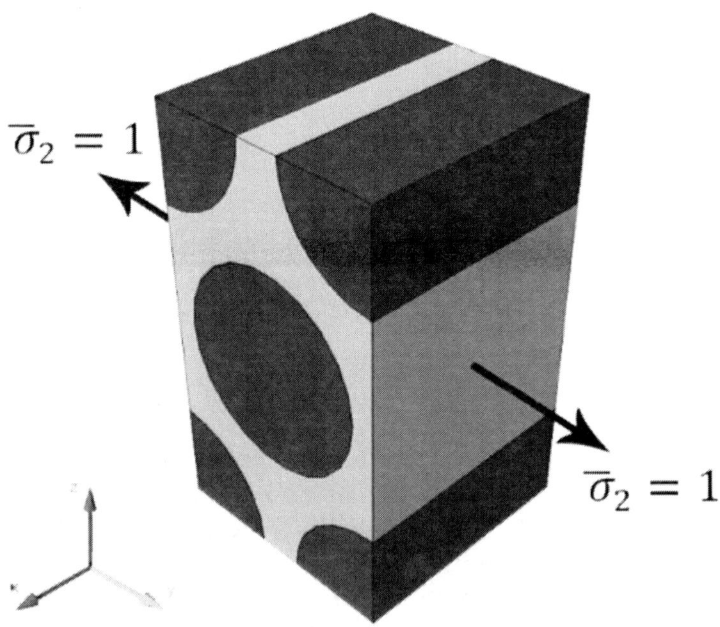

Figure 15. Fibre hexagonal array unit cell subjected to unit load.

Cesar et al. [49] reports in that there are amplification factors that relate the macroscopic ($\bar{\sigma}$)uniformly distributed unit load to the micromechanical stresses (σ) within the unit cell:

$$\sigma = M_{\sigma}\overline{\sigma} + A_{\sigma}\Delta T \quad (7)$$

$M\sigma$ and $A\sigma$ are two matrices that contain the mechanical and thermal amplification factors, respectively while ΔT represents the change in room temperature. The $M\sigma$ matrix can be calculated by applying unidirectional unit load each at a time. Therefore, for instance the first step is applying $\bar{\sigma}1=1$to get the first column of the matrix and so on. The stress amplification factor $M\sigma$ within the unit cell will vary at each point so it will end up having a contour map of the stress amplification factors over the representative volume element (RVE size: 10 mm × 5 mm). The same technique can be applied to obtain the strain amplification factors $M\varepsilon$ and $A\varepsilon$:

$$\varepsilon = M_{\varepsilon}\overline{\varepsilon} + A_{\varepsilon}\Delta T \quad (8)$$

Further details regarding applying the boundary conditions and calculating the SAF can be found in [49,50]. After obtaining the stress amplification factors, a full description of the microscopic stress distribution within the unit cell can be determined as shown in Figure 16.

Just for clarification, only the diagonal elements of the stress amplification factor tensor ($M\sigma$) are listed below. It is clear that the maximum stress is approximately 1.6 when the external applied load on the boundary is unity. The same concept has been observed experimentally, on the meso-scale, for 3D woven composites loaded in tension using image correlation [41]. This could justify why micro-fibril breakage is detected by AE event even when the applied global stress/strain is way below the ultimate strength or failure strain of fibres on the microscale or on the mesoscale. In case of a coupon specimen tested in tension, this

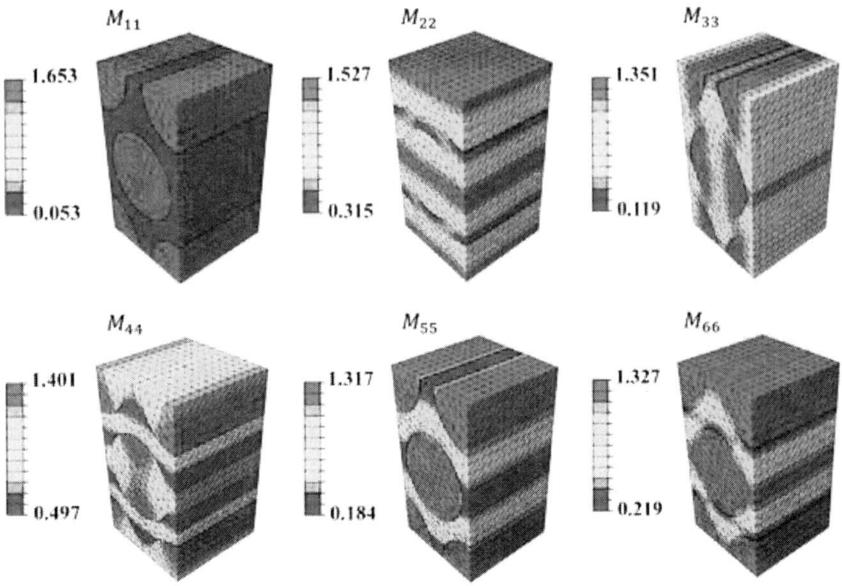

Figure 16. Diagonal elements of SAF tensor for hexagonal unit cell.

applies for the loading direction (M11) and both transverse directions (M22&M33)due to the Poisson's contraction effect; further work is required to capture more accurately the effect of the 3D fibre architecture on damage evolution.

6. Concluding Remarks

Transverse cracking in the warp yarn was detected and quantified in a 3D angle interlock woven glass composite plate during a tensile test using piezoelectric wafer active sensors bonded on the surface of the sample. The angle interlock cracking simulation have shown that the transverse component of the strain energy density is the highest when compared to the longitudinal and shear components. This implies that the strain energy release rate for the transverse component is the one that leads to transverse matrix cracking in the weft yarn under tensile loading. AE simulation has been conducted with the MP-FEM approach. The AE event was simulated as a pulse of defined duration and amplitude. The simulated electrical signal was measured at a receiver PWAS using the MP-FEM capability with the piezoelectric element. Morlet wavelet transforms and their FFT frequencies were used to process the signal in order to define and separate the different modes that composed the AE signal. These results show that the amplitude of the AE signal depends on the distance between the crack and the sensor (affected by damping). Moreover, simulated and experimental transverse cracking generates a predominant fundamental flexural mode A0 and also a less energetic fundamental extensional mode S0. Moreover, the binder yarns at the tips of the transverse crack might break which is represented by a typical AE waveform (shape and energy). This micro-fibril breakage generates only the fundamental flexural mode A0. In addition, the stress amplification factor was developed to justify why transverse matrix cracking and micro-fibril breakage is detected by AE event even when the applied global stress/strain is way below the ultimate strength or failure strain of matrix/fibres on the microscale or on the mesoscale.

In the near future, more work needs to be done on (a) calibrating the MP-FEM modelling of guided wave for accurate representation of physical phenomenon; (b) simulate the real energy release of crack growth using XFEM or VCCT model; (c) better understand the multi-modal guided wave propagation in complex 3D woven composite plates and identify more effective wave-tuning methods and signal processing algorithm for damage identification and localisation. A complete study on the guided wave propagation and the attenuation effect is also required in order to increase the accuracy of the results.

Although some good progress has been demonstrated, there are still some outstanding questions that need to be answered. A complete experimental research program and a MP-FEM method need to be fully performed in order to better understand the damage evolution (that includes multiple matrix cracks, delamination, and fibre breakage) and ultimate failure of these 3D AI glass composite plates.

Acknowledgments

All sources of funding of the study should be disclosed. Please clearly indicate grants that you have received in support of your research work. Clearly state if you received funds for covering the costs to publish in open access.

Author Contributions

Matthieu Gresil and Mohamed Nasr Saleh conceived and designed the experiments, performed the experiments, analyzed the data; Constantinos Soutis contributed to review the paper; Matthieu Gresil wrote the paper.

Conflicts of Interest

The authors declare no conflict of interest.

References

1. Gorman, M.R.; Proser, W.H. AE source by plate analysis. *J. Acoust. Emiss.* 1991, *9*, 238–288.

2. Gresil, M.; Giurgiutiu, V. Prediction of attenuated guided waves propagation in carbon fiber composites using Rayleigh damping model. *J. Intell. Mater. Syst. Struct.* **2015**, *26*, 2151–2169.

3. Mullin, J.; Berry, J.M.; Gatti, A. Some Fundamental Fracture Mechanisms Applicable to Advanced Filament Reinforced Composites. *J. Compos. Mater.* **1968**, *2*, 82–103.

4. Mehan, R.L.; Mullin, J.V. Analysis of Composite Failure Mechanisms Using Acoustic Emissions. *J. Compos. Mater.* **1971**, *5*, 266–269.

5. Godin, N.; Huguet, S.; Gaertner, R. Integration of the Kohonen's self-organising map and k-means algorithm for the segmentation of the AE data collected during tensile tests on cross-ply composites. *NDT E Int.* **2005**, *38*, 299–309.

6. Wadim, J.R. *Acoustic Emission Applications*; D. Endevco: San Juan Capistrano, CA, USA, 1978.

7. Chen, O.; Karandikar, P.; Takeda, N.; Kishi Rcast, T. Acoustic emission characterization of a glass-matrix composite. *Nondestr. Test. Eval.* **1992**, *8*, 869–878.

8. Ceysson, O.; Salvia, M.; Vincent, L. Damage mechanisms characterisation of carbon fibre/epoxy composite laminates by both electrical resistance measurements and acoustic emission analysis.*Scr. Mater.* **1996**, *34*, 1273–1280.

9. Kim, S.-T.; Lee, Y.-T. Characteristics of damage and fracture process of carbon fiber reinforced plastic under loading-unloading test by using AE method. *Mater. Sci. Eng. A* **1997**, *234*, 322–326.

10. Kotsikos, G.; Evans, J.T.; Gibson, A.G.; Hale, J. Use of acoustic emission to characterize corrosion fatigue damage accumulation in glass fiber reinforced polyester laminates. *Polym. Compos.* **1999**,*20*, 689–696.

11. Gong, X.L.; Laksimi, A.; Benzeggagh, M.L. Nouvelle approche de l'émission acoustique et son application à l'identification des mécanismes d'endommagement dans les matériaux composites.*Rev. Compos. Matér. Av.* **1998**, *8*, 179–205.

12. Masmoudi, S.; El Mahi, A.; El Guerjouma, R. Mechanical behaviour and health monitoring by acoustic emission of sandwich composite integrated by piezoelectric implant. *Compos. Part B Eng.***2014**, *67*, 76–83.

13. Barré, S.; Benzeggagh, M.L. On the use of acoustic emission to investigate damage mechanisms in glass-fibre-reinforced polypropylene. *Compos. Sci. Technol.* **1994**, *52*, 369–376.

14. Komai, K.; Minoshima, K.; Shibutani, T. Investigations of the Fracture Mechanism of Carbon/Epoxy Composites by AE Signal Analyses. *JSME Int. J.* **1991**, *34*, 381–388.

15. Li, L.; Lomov, S.V.; Yan, X.; Carvelli, V. Cluster analysis of acoustic emission signals for 2D and 3D woven glass/epoxy composites. *Compos. Struct.* **2014**, *116*, 286–299.

16. Liu, P.F.; Chu, J.K.; Liu, Y.L.; Zheng, J.Y. A study on the failure mechanisms of carbon fiber/epoxy composite laminates using acoustic emission. *Mater. Des.* **2012**, *37*, 228–235.

17. Russel, S.S.; Henneke, E.G. *Signature Analysis of Acoustic Emission from Graphite/Epoxy Composites*; Interim Report, NASA Grant NSG 1238, Report No. VPI-E-77–221977; Langley Research Center: Hampton, VA, USA, 1977.

18. Suzuki, M.; Nakanishi, H.; Iwamoto, M.; Jiao, G.-Q.; Koike, K.; Imura, M. Fatigue Fracture Mechanism of Class A-SMC by Acoustic Emission Method. *J. Soc. Mater. Sci. Jpn.* **1987**, *36*, 1402–1408.

19. Suzuki, M.; Kida, S.; Shimbo, M.; Miyano, Y. Effects of fiber content on fracture mechanisms of short fiber reinforced PET composites. *Trans. Jpn. Soc. Mech. Eng. Ser. A* **1990**, *56*, 1030–1035.

20. De Groot, P.J.; Wijnen, P.A.M.; Janssen, R.B.F. Real-time frequency determination of acoustic emission for different fracture mechanisms in carbon/epoxy composites. *Compos. Sci. Technol.***1995**, *55*, 405–412.

21. Ramirez-Jimenez, C.R.; Papadakis, N.; Reynolds, N.; Gan, T.H.; Purnell, P.; Pharaoh, M. Identification of failure modes in glass/polypropylene composites by means of the

primary frequency content of the acoustic emission event. *Compos. Sci. Technol.* **2004**, *64*, 1819–1827.

22. Gutkin, R.; Green, C.J.; Vangrattanachai, S.; Pinho, S.T.; Robinson, P.; Curtis, P.T. On acoustic emission for failure investigation in CFRP: Pattern recognition and peak frequency analyses.*Mech. Syst. Sign. Process.* **2011**, *25*, 1393–1407.

23. Bussiba, A.; Kupiec, M.; Ifergane, S.; Piat, R.; Böhlke, T. Damage evolution and fracture events sequence in various composites by acoustic emission technique. *Compos. Sci. Technol.* **2008**, *68*, 1144–1155.

24. Jong, H.-J. Transverse Cracking in a Cross-ply Composite Laminate—Detection in Acoustic Emission and Source Characterization. *J. Compos. Mater.* **2006**, *40*, 37–69.

25. Lomov, S.; Karahan, M.; Bogdanovich, A.; Verpoest, I. Monitoring of acoustic emission damage during tensile loading of 3D woven carbon/epoxy composites. *Text. Res. J.* **2014**, *84*, 1373–1384.

26. Viktorov, I.A. *Rayleigh and Lamb Waves—Physical Theory and Application*; New York Plenum Press: New York, NY, USA, 1967.

27. Achenbach, J.D. *Wave Propagation in Elastic Solids: North-Holland Series in Applied Mathematics and Mechanics*; Elsevier: Amsterdam, The Netherlands, 1973.

28. Graff, K.F. *Wave Motion in Elastic Solids*; Dover Publications: London, UK, 1975.

29. Rose, J.L. *Ultrasonic Waves in Solid Media*; Cambridge University Press: Cambridge, UK, 1999.

30. Dieulesaint, E.; Royer, D. *Ondes Élastiques Dans Les Solides—Tome 1: Propagation Libre et Guidée*; Masson: Paris, France, 1996.

31. Harker, A.H. *Elastic Waves in Solids*; British Gas/Adam Hilger: Bristol, UK, 1987. Wilcox, P.; Lowe, M.; Cawley, P. The effect of dispersion on long-range inspection using ultrasonic guided waves. *NDT E Int.* **2001**, *34*, 1–9.

32. Alleyne, D.N.; Cawley, P. The interaction of Lamb waves with defects. *IEEE Trans. Ultrason.* **1992**,*39*, 381–397.

33. Mal, A.K.; Chang, Z. Scattering of lamb wave from a rivet hole with edge cracks. *Mech. Mater.* **1999**, *31*, 197–204.

34. Lemistre, M.; Balageas, D. Structural health monitoring system based on diffracted Lamb wave analysis by multiresolution processing. *Smart Mater. Struct.* **2001**, *10*, 504–511.

35. *Standard Test Method for Tensile Properties of Polymer Matrix Composite Materials*; ASTM D3039/D3039M-14; ASTM International: West Conshohocken, PA, USA, 2014.

36. Gresil, M.; Yu, L.; Giurgiutiu, V.; Sutton, M. Predictive modeling of electromechanical impedance spectroscopy for composite materials. *Struct. Health Monit.* **2012**, *11*, 671–683.

37. Nairn, J.A. Applications of finite fracture mechanics for predicting fracture events in composites. In Proceedings of the 5th International Conference on Deformation and Fracture of Composites, London, UK, 18–19 March 1999.

38. Nasr Saleh, M.; Lubineau, G. Understanding the mechanisms that change the conductivity of damaged ITO-coated polymeric films: A micro-mechanical investigation. *Sol. Energy Mater. Sol. Cells* **2014**, *130*, 199–207.

39. Obert, E.; Daghia, F.; Ladevèze, P.; Ballere, L. Micro and meso modeling of woven composites: Transverse cracking kinetics and homogenization. *Compos. Struct.* **2014**, *117*, 212–221.

40. Saleh, M.N.; Yudhanto, A.; Potluri, P.; Lubineau, G.; Soutis, C. Characterising the loading direction sensitivity of 3D woven composites: Effect of z-binder architecture. *Compos. Part A Appl. Sci. Manuf.* **2016**. in press.

41. Cox, B.N.; Dadkhah, M.S. The Macroscopic Elasticity of 3D Woven Composites. *J. Compos. Mater.* **1995**, *29*, 785–819.

42. Yu, B.; Blanc, R.; Soutis, C.; Withers, P.J. Evolution of damage during the fatigue of 3D woven glass-fibre reinforced composites subjected to tension–tension loading observed by time-lapse X-ray tomography. *Compos. Part A Appl. Sci. Manuf.* **2016**, *82*, 279–290.

43. Yu, B.; Bradley, R.S.; Soutis, C.; Hogg, P.J.; Withers, P.J. 2D and 3D imaging of fatigue failure mechanisms of 3D woven composites. *Compos. Part A Appl. Sci. Manuf.* **2015**, *77*, 37–49.

44. Surgeon, M.; Wevers, M. Modal analysis of acoustic emission signals from CFRP laminates. *NDT E Int.* **1999**, *32*, 311–322.

45. Gresil, M.; Yu, L.; Shen, Y.; Giurgiutiu, V. Predictive model of fatigue crack detection in thick bridge steel structures with piezoelectric wafer active sensors. *Smart Struct. Syst.* **2013**, *12*, 97–119.

46. Gresil, M.; Shen, Y.; Giurgiutiu, V. Predictive modeling of ultrasonics SHM with PWAS transducers. In Proceedings of the 8th International Workshop on Structural Health Monitoring, Stanford, CA, USA, 13–15 Septmeber 2011.

47. Vallen-System GmbH, Munich, Germany. 2001. Available online: http://www.vallen.de/wavelet/index.html (accesse d on 15 January 2014).

48. Junior, P.C.P.; Bussamra, F.L.S.; Arakaki, F.K. Finite element procedure for stress amplification factor recovering in a representative volume of composite materials. *J. Aerosp. Technol. Manag.* **2011**, *3*, 239–250.

49. Mao, J.Z.; Sun, X.S.; Ridha, M.; Tan, V.B.C.; Tay, T.E. A Modeling Approach Across Length Scales for Progressive Failure Analysis of Woven Composites. *Appl. Compos. Mater.* **2012**, *20*, 213–231.

© 2016 by the authors; licensee MDPI, Basel, Switzerland. This article is an open access article distributed under the terms and conditions of the Creative Commons Attribution (CC-BY) license (http://creativecommons.org/licenses/by/4.0/).

10

CHAPTER

Computational Study of the Effect of Cortical Porosity on Ultrasound Wave Propagation in Healthy and Osteoporotic Long Bones

Vassiliki T. Potsika [1], Konstantinos N. Grivas [2], Theodoros Gortsas [2], Gianluca Iori [3], Vasilios C. Protopappas [1], Kay Raum [3], Demosthenes Polyzos [2] and Dimitrios I. Fotiadis [1,4,*]

[1]*Unit of Medical Technology and Intelligent Information Systems, Department of Materials Science and Engineering, University of Ioannina, GR 45110 Ioannina, Greece*

[2]*Department of Mechanical Engineering and Aeronautics, University of Patras, GR 26500 Patras, Greece*

[3]*Berlin-Brandenburg School for Regenerative Therapies, Charité-Universitätsmedizin Berlin, Augustenburger Platz 1, 13353 Berlin, Germany*

[4]*Foundation for Research and Technology–Hellas, Institute of Molecular Biology and Biotechnology, Department of Biomedical Research, GR 45110 Ioannina, Greece*

Abstract

Computational studies on the evaluation of bone status in cases of pathologies have gained significant interest in recent years. This work presents a parametric and systematic numerical study on ultrasound propagation in cortical bone models to investigate the effect of changes in cortical porosity and the occurrence of large basic multicellular units, simply called non-refilled resorption lacunae (RL), on the velocity of the first arriving signal (FAS). Two-dimensional geometries of cortical bone are established for various microstructural models mimicking normal and pathological tissue states. Emphasis is given on the detection of RL formation which may provoke the thinning of the cortical cortex and the increase of porosity at a later stage of the disease. The central excitation frequencies 0.5 and 1 MHz are examined. The proposed configuration consists of one point source and multiple successive receivers in order to calculate the FAS velocity in small propagation paths (local velocity) and derive a variation profile along the cortical surface. It was shown that: (a) the local FAS velocity can capture porosity changes including the occurrence of RL with different number, size and depth of formation; and (b) the excitation frequency 0.5 MHz is more sensitive for the assessment of cortical microstructure.

Keywords: ultrasound; bone modeling; osteoporosis; cortical porosity; basic multicellular units

1. Introduction

Osteoporosis is a bone pathology characterized by low bone mass and structural deterioration of bone tissue leading to bone fragility, thinning of the cortical cortex, reduced bone strength and an increased susceptibility to fractures. Nowadays, the ageing of the population has led to greater prevalence of osteoporosis which is responsible for approximately 2 million fractures which occur annually and may lead to diminished quality of life, disability and even death with a mortality rate 36% depending on the age [1,2]. In clinical practice, the golden standard for the assessment of the disease is dual energy X-ray absorptiometry scan. However, X-ray-based imaging techniques are expensive methods and require radiation exposure. Therefore, the development of non-destructive and non-radiating methods, appropriate to assess the mechanical and structural properties of bone, is crucial.

To this end, several research groups worldwide have investigated the diagnostic and monitoring role of quantitative ultrasound in osteoporotic and fractured bones using experimental, numerical and theoretical means [3,4,5,6,7,8,9]. The research interest in numerical and theoretical methods has allowed the study of phenomena which cannot be observed via experimental measurements extending thus our understanding on the underlying ultrasound propagation mechanisms and the interaction with biological tissues [10]. Specifically, as ultrasound propagates in composite media such as osseous tissues attenuation mechanisms evolve including absorption and scattering phenomena. In soft tissues the viscous forces between neighboring particles moving with different velocities are major sources of acoustic wave absorption [1]. Scattering is induced by the porous nature of cortical bone and the incident energy is partially transferred to the scattered fields. In soft tissues like blood and marrow, ultrasound bulk shear waves are usually ignored as shear waves are highly attenuated at ultrasonic frequencies, while in hard osseous tissues both compression and shear waves are considered.

In addition, the availability of advanced computer-aided computational tools in combination with imaging data derived from micro-computed tomography (μCT) and scanning acoustic microscopy (SAM) have accelerated the evolution of the numerical research [2,3,7,8]. The axial and through transmission techniques have been mainly used to estimate the first arriving signal (FAS) velocity and attenuation as well as to study the propagation of guided waves. In the literature, different thresholds for the detection of the FAS and methodologies to extract guided wave features have been proposed to interpret ultrasound propagation phenomena occurring not only at the periosteal region, but also at deeper bone layers [4,5,6,7,11,12]. More recently, ultrasound reflection and backscattering parameters have been measured to investigate bone microstructure [13,14].

Concerning the first two-dimensional (2D) computational studies, the cortical bone was modeled as a linear elastic homogeneous plate and ultrasound simulations were performed for different plate thicknesses to investigate the thinning of the cortical cortex due to osteoporosis [15,16]. In [17], X-ray computed tomography reconstructions were derived from the human radius in order to develop more realistic 3D computational models of cortical bone. It was shown that for plate thicknesses larger than the wavelength the FAS corresponds to a lateral wave which propagates at the bulk longitudinal velocity of bone, while for very thin plates the FAS wave propagates as the lowest-order symmetric plate mode. More recent 2D and 3D computational studies [2,8,14,17,18,19] take into consideration cortical microstructure, porosity and anisotropy. In [18], SAM images of human femoral neck were used to develop numerical models which account for the sample's overall shape, microstructure, cortical porosity, heterogeneous matrix elasticity and density. It was found that the FAS velocity is not influenced by trabecular bone properties or by the heterogeneities of the cortical bone mineralized matrix. On the other hand, the FAS was sensitive to variations in cortical porosity. In addition, in [2], SAM human femoral neck cross-sections were used to develop multivariate models for the prediction of pore size, porosity, and cortical thickness. It was shown that the FAS velocity decreases with increasing porosity, while an increase of the cortical thickness and pores' diameter leads to an increase in FAS velocity. It was also reported that cortical porosity has the strongest effect on the prediction of the FAS velocity. In addition, a 3D micromechanical model consisting of an anisotropic matrix pervaded by cylindrical pores was developed in [19]. It was found that, for the elderly population, the elastic properties of the mineralized matrix do not undergo large variations among different samples, while changes in the intracortical porosity account for most of the variations of mesoscopic elasticity [19].

In this work, we present a 2D computational study of ultrasonic propagation on healthy and osteoporotic cortical bone to examine the effect of porosity on the FAS velocity. We account for the occurrence of large basic multicellular units, simply called non-refilled resorption lacunae (RL), as the predominance of resorption causes an increase in Haversian canal size leading to the increase of cortical porosity and fragility [14,20]. Initially, various microstructural models are established, mimicking normal and pathological tissue states with porosity of 5%, 10% and 16%. Then, numerical models are presented to investigate the impact of changes in the number, size, position and depth of the RL on the FAS. The central excitation frequencies 0.5 and 1 MHz are examined. To our knowledge, this is the first systematic and parametric numerical study which evaluates the FAS velocity variation along cortical bone derived from small, serial propagation paths for different porosities and frequencies, including the occurrence of pores with different sizes.

2. Validation of the Numerical Method: A Benchmark Problem

The most popular numerical methods for ultrasonic evaluation of cortical bone are the Finite Difference, Finite Element and Boundary Element methods which solve partial differential equations. According to [21], for time domain modeling of broadband or high-frequency waves these methods can become cumbersome and slow due to the requirements for many grid points per wavelength and small time-steps to minimize unwanted numerical dispersion. In order to validate the finite-difference time-domain (FDTD) code that is used in this work (SimSonic, LIP, Paris, France [22,23]), a benchmark problem was established. We aim to compare the outcome of the FDTD method with a MATLAB (Version, The Mathworks, Inc., Natick, MA, USA) toolbox which combines pseudo-spectral and k-space methods, so the spatial gradients are calculated by using Fast Fourier Transforms rather than by using a finite difference stencil (k-Wave, [21]). The computational model of Figure 1a was used corresponding to an intact cortical plate with thickness of 4 mm and length 40 mm surrounded by water. The material properties of bone and water are presented in Table 1. The grid size was set to 20 μm and the wave propagation simulation time was 10 μs. A point source and receiver were used to calculate the velocity at a distance of 12.5 mm between the transducers. The transducers were placed directly onto the cortical surface. A 1-MHz Gaussian pulse was used as the excitation signal. According to the results of Figure 2, the two waveforms have the same time of arrival of the FAS as well as the same qualititative behavior over time enhancing the accuracy of the FDTD method.

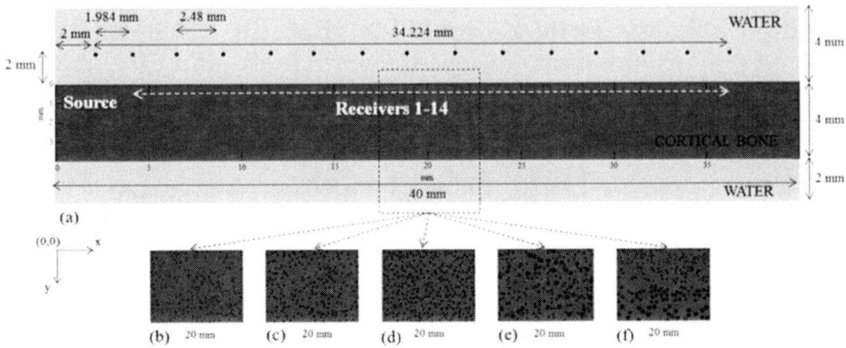

Figure 1. Numerical models of cortical bone corresponding to the first set of simulations (Series I_Po0-16) and Table 2, namely: (**a**) B_Hom; (**b**) B_Po5; (**c**) B_Po10; (**d**) B_Po16; (**e**) B_Po16_RL; and (**f**) B_Po16_Gradual. For each geometry, only one of the three random porosity distributions is depicted in the form of 5 mm cortical segments derived from the original 40mm × 4mm plates. The ultrasound configuration is also presented in (**a**).

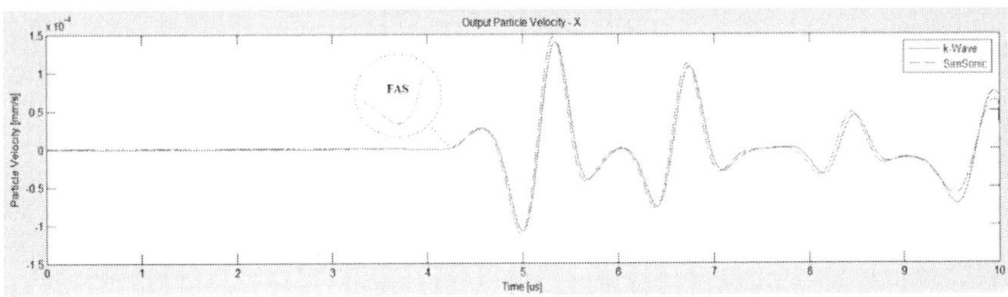

Figure 2. Particle velocity waveforms derived from two different numerical tools, Simsonic and k-Wave, for grid size 20 μm. The threshold for the detection of the FAS is also depicted.

Table 1. Elastic coefficients and density used for the simulations [14].

Propagation Medium	C_{11} (GPa)	C_{12} (GPa)	C_{66} (GPa)	P (g/cm^3)
Water	2.25	2.25	0	1.00
Bone	28.71	10.67	9.02	1.85

In order to provide quantitative information for the differences of the two waveforms, a convergence study is conducted for the grid sizes of 50, 30 and 20 μm. The methodology of waveform rectification was used followed by numerical integration via the trapezoidal method. A full-wave rectifier converts the examined waveform to a waveform of constant polarity (positive or negative). Specifically, the excitation signal was rectified as well as the received waveforms from k-Wave and Simsonic. Numerical integration was applied using the trapezoidal method to derive the emitted and received signal area. Next, we calculated the ratio of the emitted signal area to the received waveforms' area, which is a quantitative indicator reflecting the whole area of interest. If we denote the calculated ratio with E_{sim} for Simsonic and E_{kw} for k-Wave, the corresponding values were: (a) E_{kw} = 1166, E_{sim} = 1084 ($\text{difference}_{50\mu m}$= 82) for 50 μm, (b) E_{kw} = 1935, E_{sim} = 1921 ($\text{difference}_{30\mu m}$=14) for 30 μm and (c) E_{kw} = 2924, E_{sim} = 2919 ($\text{difference}_{20\mu m}$ = 5) for 20 μm.

3. Numerical Evaluation of Cortical Porosity Using Ultrasonic Techniques

3.1. Model Geometry

In this section, we present the structural features of the computational models representing different cases of cortical porosity and pores' sizes as well as different number and distributions of RL. The structural and material properties were derived from [14], in which the Haversian canals of normal size were differentiated from the RLs to distinguish healthy bones (with no or few RL) from degenerated ones. As can be seen in Figure 1, the cortical bone was modeled as a 2D plate (length 40 mm, width 4 mm) surrounded by water (upper surface 4 mm, lower surface 2 mm).

Two sets of numerical simulations were performed in order to evaluate: (a) porosity changes from 0% to 16% covering the whole region of cortical bone (Series I_Po0-16), (b) the occurrence of a single or a cluster of RL in small propagation paths (Series II_RL). The selection of the porosities of Series I_Po0-16, and the sizes of the RL were based on previous studies investigating the interaction of ultrasound with cortical porosity [2,5,10,14]. However, it should be emphasized that the evaluation of cortical porosity depends on several parameters such as age, gender and region of interest [5]. The first set is presented in Figure 1, in which cortical bone was modeled both as a homogeneous and nonhomogeneous medium to account for the cases of: (a) intact bone (Figure 1a); (b) porosity 5% (Figure 1b); (c) porosity 10% (Figure 1c); and (d) porosity 16% (Figure 1d–f). In Table 2, we present the exact structural features concerning the porosity, number and dimensions of the normal and larger pores. For each microstructural scenario of Table 2, a total of three numerical models were established and the mean local FAS velocity values were calculated as well as the standard error.

Table 2. Examined cases in the tangential direction for the second set of the simulations.

	Description	Porosity (%)	No. of Pores	Pores' Radius (μm)	No. of RL/RLs	Radius of RL/RLs (μm)
B_Hom	Homogeneous bone	0	0	–	–	–
B_Po5	Porous bone, normal pores	5	1592	40	–	–
B_Po5_RL1	Porous bone, normal pores and 1 RL	5	1592	40	1	115
B_Po5_RL3	Porous bone, normal pores and 3 RLs	5	1592	40	3	115
B_Po5_RL5	Porous bone, normal pores and 5 RLs	5	1592	40	5	115
B_Po10	Porous bone, normal pores	10	1592	60	–	–
B_Po16	Porous bone, normal pores	16	2263	60	–	–
B_Po16_RL	Porous bone, normal pores and RLs	16	1400	60	235	115
B_Po16_Gradual	Gradual distribution of the pores, normal pores and RLs	16	1775	40, 60, 80	192	115

Circular scatterers were considered in the transverse direction and the selected radii of normal pores and RL are presented in Table 2. Given the porosity and the dimensions of the plate and scatterers, the number of the pores can be calculated. An algorithm was developed to generate the osseous and soft tissues as greyscale values and the "porosity maps" with a random distribution of the pores. The scatterers were not in contact and did not merge with each other as well as with the cortical boundaries.

Figure 3 illustrates cortical segments of Series II_RL in which the cortex of a long bone was modeled initially as a 2D homogeneous plate including the presence of 1 RL at different lengths, depths and diameters (Figure 3a–e). Then, in Figure 3f–h, we account for cortical microstructure to simulate the case of normal remodeling (Table 2: cases B_Po5_RL1, B_Po5_RL3, B_Po5_RL5) including a single or a cluster of RL in the center of the plate.

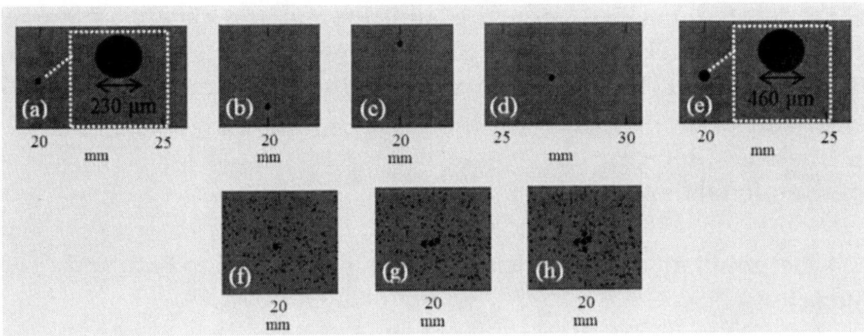

Figure 3. Numerical models of cortical bone corresponding to the second set of simulations (Series II_RL) focusing on the regions of the occurrence of RL: (**a**) homogeneous bone, 1 RL with center coordinates (x, y) = (20, 2); (**b**) homogeneous bone, 1 RL with center coordinates (x, y) = (20, 3); (**c**) homogeneous bone, 1 RL with center coordinates (x, y) = (20, 1); (**d**) homogeneous bone, 1 RL with center coordinates (x, y) = (27, 2) mm; (**e**) homogeneous bone, 1 RL with center coordinates (x, y) = (20, 2) and double the diameter; (**f**) B_Po5_RL1; (**g**) B_Po5_RL3; (**h**) B_Po5_RL5.

3.2. Material Properties

The cortical bone was modeled as an isotropic, linear elastic medium. We assumed that the soft tissues surrounding the cortical plate as well as the circular pores are composed of water. Table 1 summarizes the material properties assigned to the cortical bone, scatterers and soft tissues which were derived from [14].

3.3. Ultrasound Configuration

Simulations of ultrasound propagation were performed in the tangential direction (Figure 1a). Calculations were conducted by placing one point source and 14 point receivers: (a) at a distance 2 mm from the upper cortical layer (non-contact transducers); (b) directly onto the cortical cortex (implanted transducers). The distance between two successive receivers was 2.48 mm implying that the distance between the emitter and the receivers ranges from 1.984 to 34.224 mm (receiver 1 (R1) – receiver 14 (R14)). The width of each transducer was equal to 11 elements. The use of a linear array of transducers is in agreement with exsisting ultrasonic devices for bone characterization such as the bidirectional device in [24].

3.4. Determination of the Ultrasonic Wave Propagation Path and Velocity

The "local FAS velocities" correspond to small propagation paths and were calculated as the difference in the FAS arrival time of the signals from successive receivers (distance 2.48 mm). In this way, we intend to estimate the structural changes along small regions of cortical bone due to porosity variations or to the presence of larger pores. When multiple receivers are considered, the change in the FAS arrival time Δt (μs) can be calculated using $\Delta t(x) = t(x_i) - t(x_{i-1})$, where i is the number of the receiving position, $t(x_i)$ is the arrival time of the signal at receiver i and $t(x_{i-1})$ is the arrival time at the previous receiver [25]. For the detection of the FAS a threshold was applied to the receiving waveforms corresponding to the identification of the first signal extremum (Figure 2) [26].

3.5. Boundary Conditions

The cortical cortex was surrounded by soft tissues and perfectly matched layers (PML) were used to limit spurious reflections from the boundaries simulating an infinitely long model. The PML efficiency was set to 80 dB which means that the wave reflected by the PML is expected to be 80 dB below the amplitude of a incident wave, in the case of normal incidence [22,23].

3.6. Numerical Simulation and Signal Analysis in the Time Domain

The numerical solution of the 2D wave propagation problem was carried out by using an FDTD code [22,23]. The computations are based on a system of elastodynamic equations corresponding to the propagation of mechanical waves in continuous media which obey Hooke's law and are expressed as:

$$\rho(x)\frac{\partial u_i}{\partial t}(x,t)=\sum_{j=1}^{d}\frac{\partial T_{ij}}{\partial x_j}(x,t)+f_i(x,t) \quad (1)$$

$$\frac{\partial T_{ij}}{\partial t}(x,t)=\sum_{k=1}^{d}\sum_{l=1}^{d}c_{ijkl}(x)\frac{\partial u_k}{\partial x_l}(x,t)+\theta_{ij}(x,t) \quad (2)$$

where subscripts i={1,...,d} refer to the direction of space and d to the space dimension (d =2 for 2D computational models), x and t are the space and time variables, $\rho(x)$ is the mass density and c(x) is the fourth-order rigidity tensor. These parameters entirely define the material properties and geometry of the medium. u_i (x, t) are the vector components of the particle velocity field, T_{ij} (x, t) are the components of the stress tensor, while f_i denote the vector components of force sources and θ_{ij} denote the tensor components of strain rate sources.

A Hanning pulse was used as the excitation signal including four sinusoidal cycles [27]. The examined central angular frequencies were 0.5 and 1 MHz. The duration of the simulations was 25 μs. The accuracy of the solution depends on the relation between the element size and the wavelength. According to the software requirements, the stability condition is defined as:

$$\Delta t \leqslant \frac{1}{\sqrt{d}}\frac{\Delta x}{c_{max}} \quad (3)$$

where Δt and Δx denote the time and spatial steps used to approximate time or spatial derivatives, c_{max} is the largest speed of sound amongst all the simulated materials ($c_{max} = c_{bone} =$ 3939 m/s), and d refers to the space dimension ($d = 2$). The grid size was set to 8 μm, and a convergence study is presented in the following subsection to demonstrate the accuracy of the findings.

Convergence Study

The grid size must be dense enough to ensure an accurate numerical solution and to minimize the computational cost. In addition, it must be small enough in comparison to the diameter of the pores. The convergence study is based on the numerical model B_Po5 which integrates the smallest scatterers among all the geometries of Table 2. Two transducers were considered which were placed directly onto the cortical surface and their distance was set to 20 mm. FDTD simulations were conducted and the FAS velocities were calculated for the central excitation frequency 1 MHz and the grid step sizes 4, 8, and 16 μm. The calculated FAS values were 3499 m/s for 4 μm, 3495 m/s for 8 μm, and 3489 m/s for 16 μm. It can be concluded that the 8-μm grid size is a good compromise for an accurate numerical solution and less time consuming simulations which is in agreement with [2].

3.7. Statistical Analysis

A statistical analysis was conducted and the results are presented as local mean FAS velocity ± standard error. Linear regression analysis and one-way analysis of variance (one-way ANOVA) were used to evaluate changes in porosity. The statistical findings were considered significant for p-values less than 0.05 [9].

4. Results

Figure 4, Figure 5, Figure 6, Figure 7, Figure 8, Figure 9, Figure 10, Figure 11 and Figure 12 present the results derived from the two simulation sets of Figure 1 and Figure 3 and the central excitation frequencies 0.5 and 1 MHz. As multiple receivers were considered and small propagation paths were examined, the horizontal axis represents the distance of each receiving element from the starting point of the plate with coordinates $(x, y) = (0, 0)$. First, in Subsection 4.1, we present the results derived for porosity changes from 0% to 16%, and then, in Subsection 4.2, the findings corresponding to the occurrence of a single or a cluster of RLs in small propagation paths at different depths, lengths and sizes.

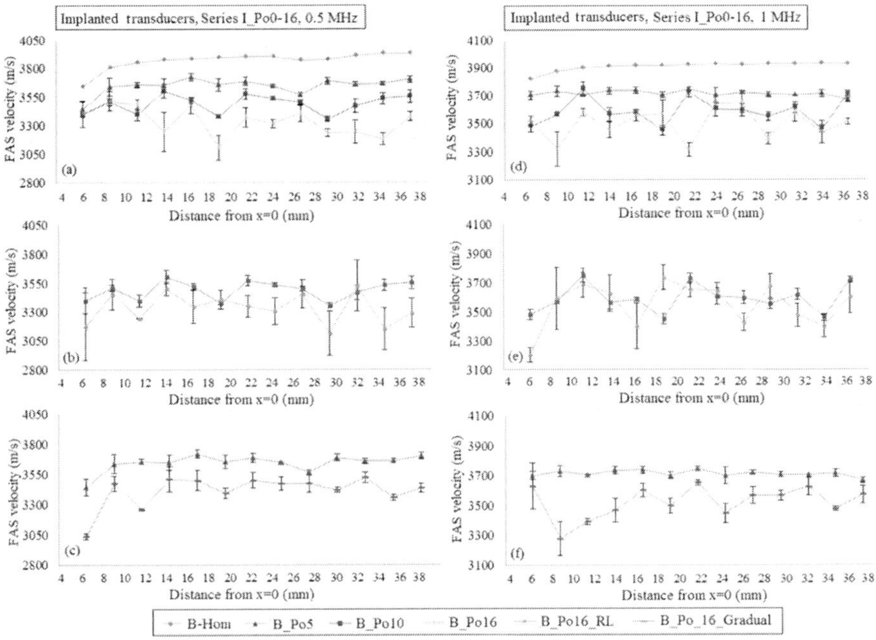

Figure 4. The potential of the mean local FAS velocity to detect changes in cortical porosity from 0%–16%. The diagrams correspond to the first set of simulations under the assumption of implanted transducers and the excitation frequencies: (**a**)–(**c**) 0.5 MHz; and (**d**)–(**f**) 1 MHz. The standard error bars demonstrate the FAS velocity variation for the three numerical models established for each porosity scenario.

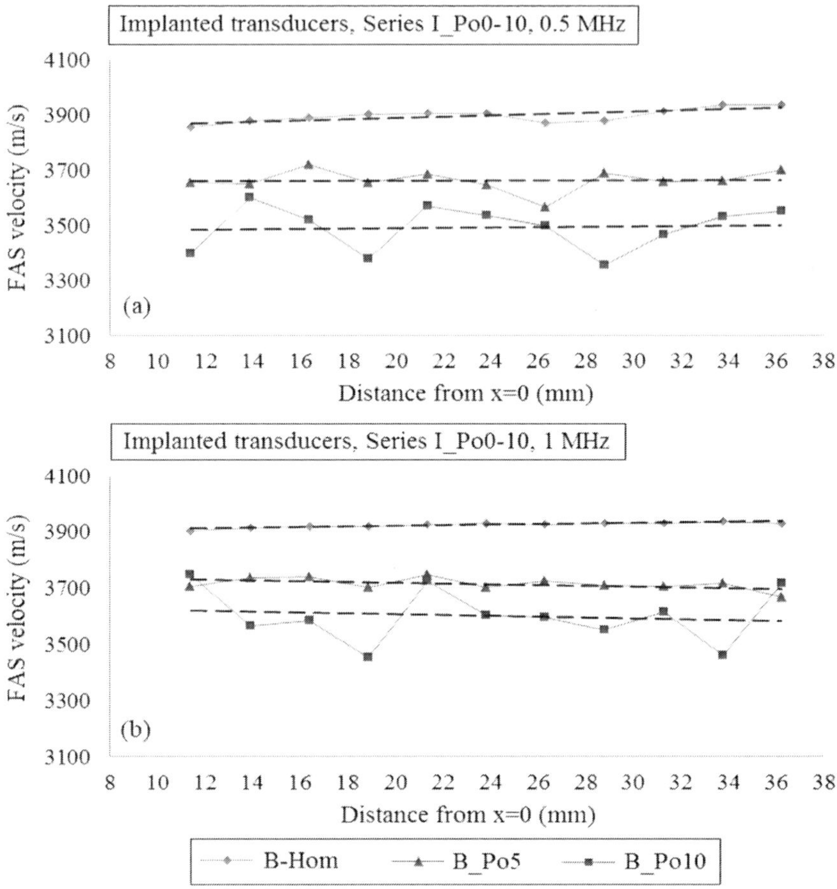

Figure 5. Linear fit derived from multiple receiving positions for cortical porosities from 0 to 10%. The diagrams correspond to the first set of simulations under the assumption of implanted transducers and the excitation frequencies: (**a**) 0.5 MHz; and (**b**) 1 MHz. The corresponding equations, coefficients of determination and root mean square errors are presented in Table 3 and Table 4.

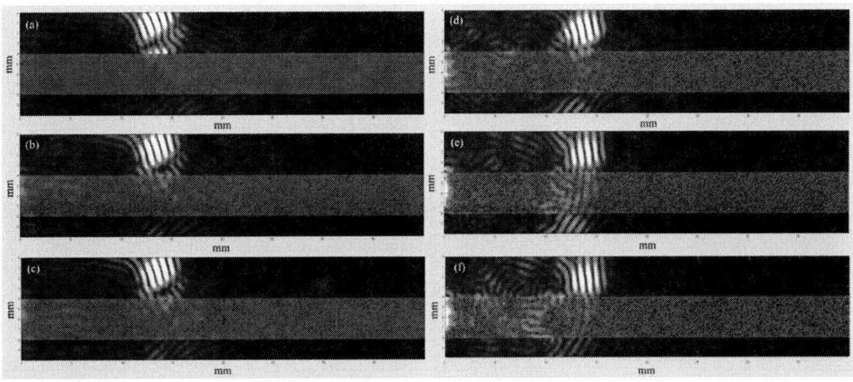

Figure 6. Snapshots of wave propagation for the first set of simulations using non-contact transducers for the excitation frequency of 1 MHz and time instant 10 μs for the cases: (**a**) B_Hom; (**b**) B_Po5; (**c**) B_Po5_RL5; (**d**) B_Po10; (**e**) B_Po16; (**f**) B_Po16_RL.

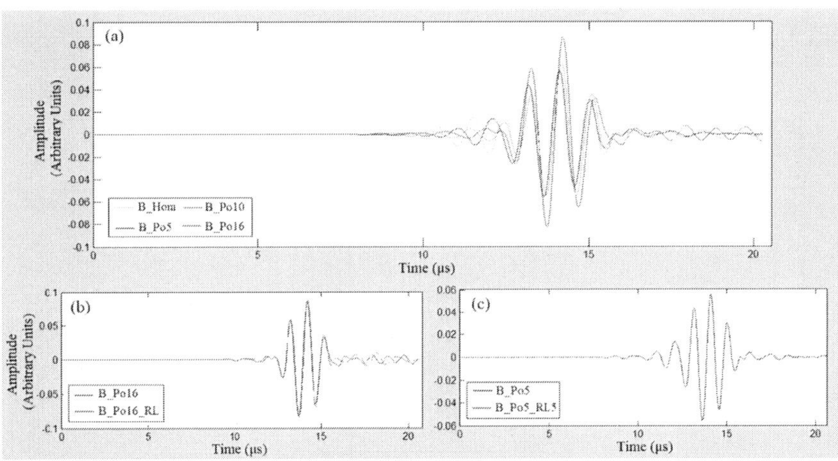

Figure 7. Waveforms derived from receiver R8 for the excitation frequency 1 MHz and simulation time 20 μscomparing the examined cases: (**a**)B_Hom, B_Po5, B_Po10 and B_Po16; (**b**) B_Po16 and B_Po16_RL; (**c**) B_Po5 and B_Po5_RL5.

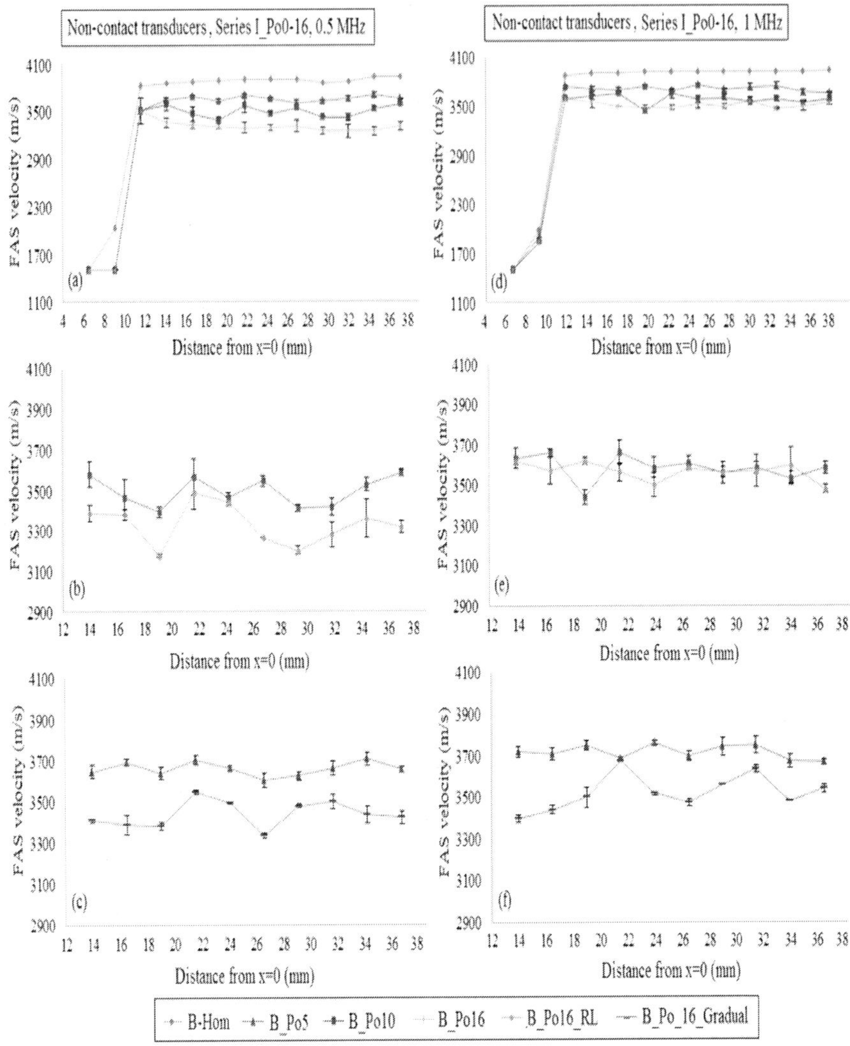

Figure 8. The potential of the mean local FAS velocity to detect changes in cortical porosity from 0% to 16%. The diagrams correspond to the first set of simulations when the transducers are placed at a distance of 2 mm from the cortical cortex. The results for the excitation frequencies : **(a)**–**(c)** 0.5 MHz; and **(d)**–**(f)** 1 MHz are presented. The standard error bars demonstrate the FAS velocity variation for the three numerical models established for each porosity scenario.

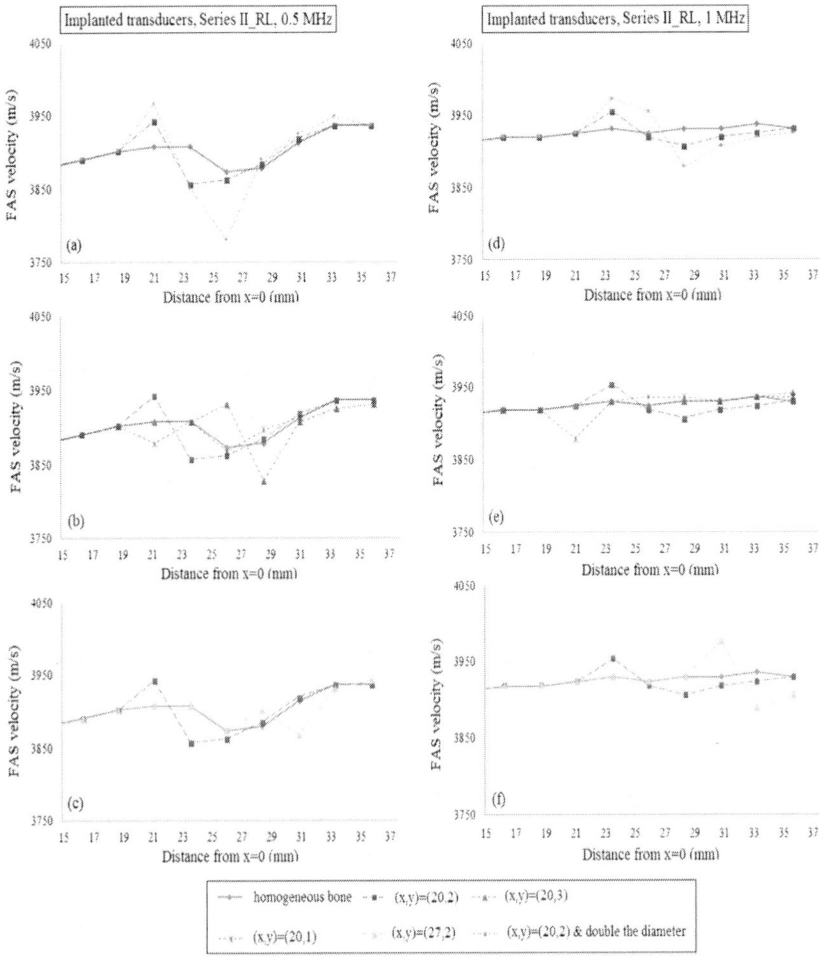

Figure 9. The potential of the local FAS velocity to detect the occurrence of a single RL considering cortical bone as a homogeneous medium (geometries illustrated in Figure 3a–e). The diagrams correspond to the second set of simulations under the assumption of implanted transducers and the excitation frequencies: (**a**)–(**c**) 0.5 MHz; and (**d**)–(**f**) 1 MHz. The legend describes the center coordinates of the RL.

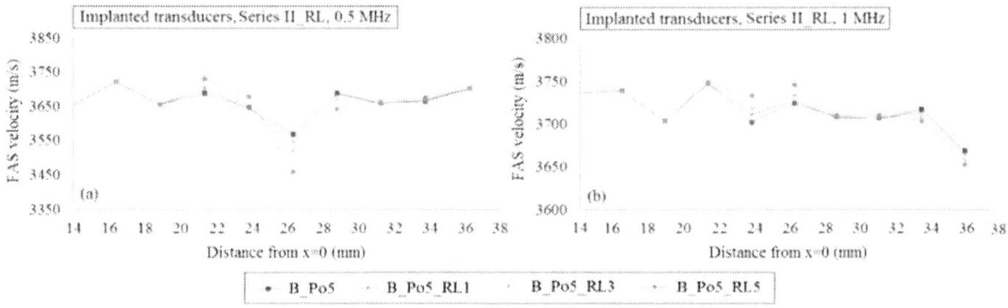

Figure 10. The potential of the local FAS velocity to detect the occurrence of a single or a cluster of RL considering cortical bone as a nonhomogeneous medium for the cases B_Po5, B_Po5_RL1, B_Po5_RL3 and B_Po5_RL5 (geometries illustrated in Figure 3f–h). The diagrams correspond to the second set of simulations under the assumption of implanted transducers and the excitation frequencies: (**a**) 0.5 MHz; and (**b**) 1 MHz.

4.1. First Set of Simulations (Series I_Po0-16)

4.1.1. Implanted Transducers

Figure 4 presents the FAS velocity variation profile from multiple implanted receivers if we consider cortical bone as a nonhomogeneous medium with porosity from 0 to 16%. The local FAS velocities were calculated as the mean values from the same receiving position of three simulation maps with different random distributions of the pores keeping the same microstructure characteristics. The mean FAS velocities and the standard error bars for the excitation frequency 0.5 MHz are presented in Figure 4a–c and for 1 MHz in Figure 4d–f. It can be observed that a porosity increase leads to the decrease of the FAS velocity variation profile, and this is more apparent for the excitation frequency 0.5 MHz.

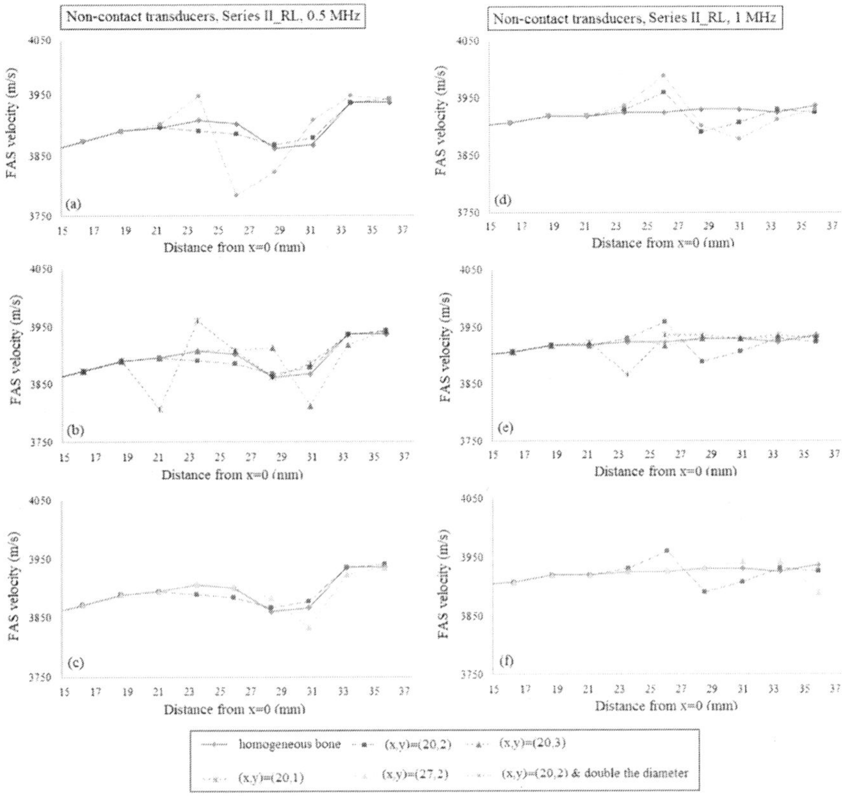

Figure 11. The potential of the local FAS velocity to detect the occurrence of a single RL considering cortical bone as a homogeneous medium (geometries illustrated in Figure 3a–e). The diagrams correspond to the second set of simulations when the transducers are placed at a distance of 2 mm from the cortical cortex. The results for the excitation frequencies : (**a**)–(**c**) 0.5 MHz; and (**d**)–(**f**) 1 MHz are presented. The legend describes the center coordinates of the RL.

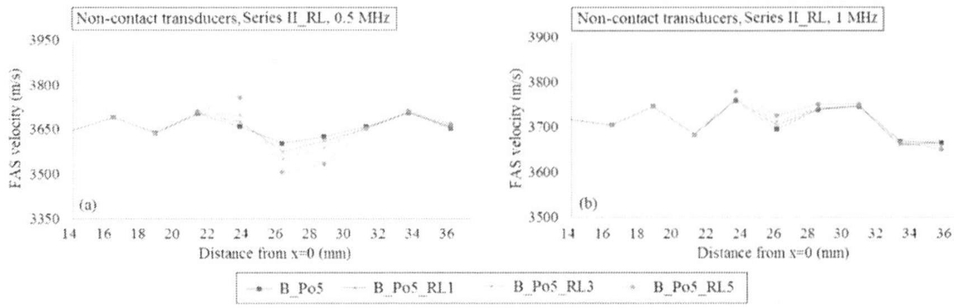

Figure 12. The potential of the local FAS velocity to detect the occurrence of a single or a cluster of RL considering cortical bone as a nonhomogeneous medium for the cases B_Po5, B_Po5_RL1, B_Po5_RL3 and B_Po5_RL5 (geometries illustrated in Figure 3f–h). The diagrams correspond to the second set of simulations when the transducers are placed at a distance of 2 mm from the cortical cortex. The excitation frequencies: (**a**) 0.5 MHz; and (**b**) 1 MHz are examined.

More specifically, if we consider the case B_Hom as the reference case, the relative percentage differences in Figure 4a for the cases B_Po5, B_Po10 and B_Po16 were calculated in the range 4.32%–7.87%, 6.58%–13.53% and 6.41%–20.34%, respectively. These findings can be directly compared withFigure 4d. For the excitation frequency 1 MHz, the relative percentage differences for the cases B_Po5, B_Po10 and B_Po16 were from 3.14% to 6.67%, 3.91%–12.10% and 7.35%–15.56%, respectively. Then,Figure 4b,c,e,f present more realistic scenarios of pathology including both normal pores and RLs for the case B_Po16_RL (Figure 4b for 0.5 MHz and Figure 4e for 1 MHz) and the gradual formation of osteoporosis (Figure 4c for 0.5 MHz and Figure 4f for 1 MHz). In Figure 4b, a porosity increase up to 16% including the occurrence of RLs (B_Po16_RL) leads to a decrease of the local FAS velocities and the relative percentage differences were from 9.23% to 19.71% for the excitation frequency 0.5 MHz. In addition, we observe that there are few positions (e.g. Figure 4b, x = 30–32 mm) which show a minor increase of the FAS velocity locally despite the porosity increase. Figure 4e presents the FAS velocities for B_Po16_RL and the excitation frequency 1 MHz. It can be seen that as the frequency increases it is more difficult to discriminate a specific FAS velocity behavior among the cases B_Po10 and B_Po16_RL. Finally, Figure 4c,f show the potential of the local FAS velocity to detect the gradual formation of osteoporosis. The diagram B_Po5 is depicted in the same diagram with B_Po16_Gradual as when osteoporosis is gradually formed the upper cortical layer has the microstructure of a healthy bone, while the porosity and pores' size increase starts from the endosteal region. The relative percentage differences for the case B_Po16_Gradual keeping B_Hom as the reference were calculated from 8.70% to 16.63% for the excitation frequency 0.5 MHz (Figure 4c) and from 5.02% to 15.49% for 1 MHz (Figure 4f), respectively.

Then, one-way ANOVA was conducted and the p-value was calculated. It was found to be lower than 0.05 ($p \approx 0$), implying that there are significant differences between group means. Linear regression analysis was also performed, as in some receiving positions of Figure 4 the mean FAS velocity was found to increase locally despite the porosity increase. Figure 5 shows that linear regression analysis is not influenced by local phenomena, and the examined cases can be discriminated. The corresponding equations, coefficients of determination (R^2) and root mean square errors (RMSE) are included in Table 3and Table 4. All the results are not presented graphically to reduce complexity, as similar findings toFigure 5 were derived.

Table 3. Linear fit of the mean first arriving signal (FAS) velocities derived from all the receiving positions and the excitation frequency 0.5 MHz.

	Implanted Transducers			Transducers at a Distance of 2 mm from the Cortical Surface		
	$y = ax + b$	R^2	RMSE	$y = ax + b$	R^2	RMSE
B_Hom	$y = 2.27x + 3844$	0.54	16.56	$y = 3.00x + 3815$	0.52	22.79
B_Po5	$y = 0.17x + 3661$	0.01	37.94	$y = 2.81x + 3577$	0.17	48.84
B_Po10	$y = 0.72x + 3475$	0.01	78.51	$y = 0.28x + 3489$	0.01	66.92
B_Po16	$y = -4.37x + 3417$	0.09	110.08	$y = -3.64x + 3385$	0.53	24.58
B_Po16_RL	$y = -4.09x + 3434$	0.06	123.45	$y = -3.23x + 3410$	0.06	92.93
B_Po16_Gradual	$y = 1.38x + 3409$	0.02	75.88	$y = 1.57x + 3397$	0.03	59.72

Table 4. Linear fit of the mean FAS velocities derived from all the receiving positions and the excitation frequency 1 MHz.

	Implanted Transducers			Transducers at a Distance of 2 mm from the Cortical Surface		
	$y = ax + b$	R^2	RMSE	$y = ax + b$	R^2	RMSE
B_Hom	$y = 1.08x + 3898$	0.80	4.26	$y = 1.65x + 3879$	0.80	6.37
B_Po5	$y = -1.41x + 3748$	0.28	17.82	$y = -1.81x + 3759$	0.19	29.16
B_Po10	$y = -1.54x + 3639$	0.02	93.23	$y = -1.78x + 3628$	0.06	57.31
B_Po16	$y = -1.36x + 3549$	0.01	100.58	$y = -0.06x + 3510$	0.01	30.84
B_Po16_RL	$y = -5.38x + 3702$	0.13	111.10	$y = -3.21x + 3647$	0.28	36.73
B_Po16_Gradual	$y = 3.96x + 3440$	0.15	72.96	$y = 4.49x + 3406$	0.16	73.61

4.1.2. Transducers at a Distance of Two Millimeters from the Cortical Cortex

Figure 6 presents snapshots of ultrasound wave propagation for indicative cases considering the excitation frequency 1 MHz and time instant 10 μs. It is shown that, as porosity increases, multiple scattering mechanisms evolve (Figure 6d,e,f). In addition, it can be observed that the application of PML limits spurious reflections from the boundaries, but the phenomenon is not totally diminished.

Then, in Figure 7 the received waveforms are illustrated derived from receiver R8, which is placed above the region of the cluster of 5 RLs (Figure 3h) for indicative cases and the excitation frequency 1 MHz, It can be seen in Figure 7a that the increase of porosity induces a delay in the propagation of the signal if we compare the zero-crossing times. An increase in amplitude was also found with increasing the porosity. A comparison of the waveforms of the cases B_Po16 and B_Po16_RL is presented in Figure 7b, showing that the occurrence of the larger pores has an effect on the received waveforms. However, in Figure 7c, the waveforms for the cases B_Po5 and B_Po5_RL5 almost coincide, while no difference can be observed at the corresponding snapshots of Figure 6.

Figure 8 presents the mean FAS velocities and standard error bars when the transducers are placed at a distance of 2 mm from the cortical surface. The porosities from 0% to 16% are examined as in Figure 4. In Figure 8a,d, low FAS velocities are observed for the receivers which are placed before $x = 12$ mm due to the direct propagation of the FAS via the soft tissues.

More specifically, if we consider the case B_Hom as the reference case, the relative percentage differences in Figure 8a for 0.5 MHz and B_Po5, B_Po10 and B_Po16 were calculated in the range 4.71%–7.65%, 7.22%–12.83% and 12.90%–17.60%,respectively. For the excitation frequency 1 MHz (Figure 8d), the relative percentage differences for B_Po5, B_Po10 and B_Po16 were from 4.24% to 6.89%, 6.37%–12.19% and 8.78%–11.82%, respectively. Then, Figure 8b,e corresponds to the case B_Po16_RL and are directly compared with the case B_Po10. Keeping B_Hom as the reference case the relative percentage differences for B_Po16_RL were from 10.51 to 18.33% for 0.5 MHz (Figure 8b) and 7.17%–11.62% for 1 MHz (Figure 8e). Thus, even if according to Figure 7b, the two waveforms seem to coincide for propagation time less then 10 μs, however the calculation of the FAS in small propagation paths shows quantitative differences when large pores are considered. The results for the gradual distribution of the pores are shown in Figure 8c,f. The relative percentage differences for B_Po16_Gradual were derived from 9.02% to 14.55% for the excitation frequency 0.5 MHz (Figure 8c) and from 6.31% to 12.95% for 1 MHz (Figure 8f), respectively. It should be mentioned that in Figure 8b,c,e,f, we do not present the whole propagation path, as the initial low values of Figure 8a,b corresponding to the direct FAS propagation through the water were neglected. In addition, linear regression analysis was performed and the results are depicted in Table 3 and Table 4. It was shown that the examined cases can be discriminated even for the diagrams of Figure 8e in which the mean local FAS velocities do not show a specific behavior with increasing the porosity.

Concerning one-way ANOVA, the p-value was lower than 0.05 ($p \approx 0$) ignoring the receiving positions near the source ($d < 12$ mm). The reason for not considering the first receivers in the calculations is that as it can be seen in Figure 8a,d for $d < 12$ mm, the FAS wave propagates directly in the soft tissues and low velocities are derived in comparison to the bulk longitudinal velocity of bone. This assumption was also considered for the cases of Figure 4 to keep the same conditions.

4.2. Second Set of Simulations (Series II_RL)

4.2.1. Implanted Transducers

This section presents the results corresponding to the occurrence of: (a) a single RL at different depths, lengths and sizes under the assumption of cortical homogeneous bone (Figure 9 and Figure 11); (b) a single or a cluster of RL taking into account cortical microstructure (Figure 10 and Figure 12). InFigure 9 and Figure 10, implanted transducers are used and the excitation frequencies 0.5 and 1 MHz are examined. Figure 9a–c depict the local FAS velocity variation profile for the excitation frequency 0.5 MHz and the presence of a single RL. For comparison purposes, all the diagrams include the calculated values for the intact bone (blue line) and the case of a single RL with center coordinates (x, y) = (20, 2) (red line). All the diagrams start from a distance of 15 mm in the x axis as for lower cortical lengths all the curves coincide. In Figure 9a, we observe that for 0.5 MHz the occurrence of one RL in the center of the plate can be captured by the receivers which are placed directly above the region of interest showing an increase of the FAS velocity of 35 m/s compared to the homogeneous bone diagram. In addition, if we keep the same RL position and double the diameter of the pore (orange line), the receivers which are placed in the position x = 19–21 mm are the first to detect an increase of the FAS velocity of 58 m/s in comparison to the homogeneous bone value. In Figure 9b, we examine how the depth of the RL occurrence influences the FAS velocity. It can be seen that, if we place the RL at a higher cortical surface (grey line), the receivers which first capture the presence of a pore are placed again in the propagation path between x = 19–21 mm showing a velocity decrease of 29 m/s comparing to the reference case of homogeneous bone. On the other hand, if we increase the depth from the cortical surface (purple line), there is a delay in the identification of the RL by the receivers above the region of x = 25–21 mm showing an increase of the FAS velocity of 57 m/s. Finally, Figure 9c shows that if we change the RL position along the x axis, the receivers which are placed above the region of interest are the first to detect an increase of the FAS velocity of 23 m/s revealing the position of the RL formation.

Figure 9d–f presents the local FAS velocity variation profile for the excitation frequency 1 MHz and the presence of a single RL. In Figure 9d, it can be seen that the occurrence of a single RL in the center of the plate (red line) can be identified with a delay by the receivers which follow the region of interest (around x = 23 mm) showing an increase of the FAS velocity of 23 m/s. In addition, it was found that if we double the diameter of the RL (orange line), the same receivers are the first to detect an increase of the FAS velocity of 41 m/s. Then, Figure 9e indicates that by increasing the depth from the cortical surface (purple line), the larger pore cannot be detected as this diagram coincides with the case of homogeneous bone (blue line). On the other hand, the occurrence of the RL near the upper cortical surface (grey line) leads to a decrease of the FAS velocity of 46 m/s derived from the receivers directly above the region of interest. Finally, Figure 9f shows that if we change the position of the RL along the $x$axis, the receivers which are placed approximately at a distance of 3 mm after the pore are the first to capture an increase of the FAS velocity of 47 m/s.

Then, in Figure 10, we investigate the potential of the local FAS velocity to detect the occurrence of a single or a cluster of three or five large pores in the center of the plate taking into account the nonhomogeneous nature of cortical bone. Figure 10a presents the FAS velocity variation profile for the excitation frequency of 0.5 MHz, while Figure 10b the FAS values for 1 MHz considering the diagram of B_Po5 (red line) as the reference case. For the excitation frequency of 0.5 MHz the relative percentage differences were calculated in the range: (a) 0.04%–0.28% for B_Po5_RL1, (b) 0.04%–1.36% for B_Po5_RL3 and (c) 0.04%–3.08% for B_Po5_RL5. It can be seen that the first receivers which can identify the RL occurrence are placed above the region of interest (x = 20–22 mm). A 2-mm delay in the detection of the larger pores is observed in Figure 10b for 1 MHz as they are first identified from the receivers in the region x = 22–24 mm. Lower relative percentage differences were calculated compared to 0.5 MHz from: (a) 0.05%–0.24% for B_Po5_RL1, (b), 0.23%–0.42% for B_Po5_RL3, and (c) 0.05%–0.85% for B_Po5_RL5.

4.2.2. Transducers at a Distance of Two Millimeters from the Cortical Cortex

Figure 11a–c present the local FAS velocity variation profile for the excitation frequency 0.5 MHz and the presence of a single RL by placing the transducers at a distance of 2 mm from the cortical cortex. All the diagrams include the case of intact homogeneous bone (blue line) and the case of a single RL with center coordinates $(x, y) = (20, 2)$ (red line). In Figure 11a, it can be seen that, for 0.5 MHz, the occurrence of one RL in the center of the plate can be captured by the receivers which are placed directly after the region of interest (between x = 21–23 mm) showing a decrease of the FAS velocity of 17 m/s compared to the homogeneous bone diagram. Then, if we keep the same RL position and double the diameter of the RL (orange line) the same receivers are the first to detect an increase of the FAS velocity of 41 m/s in comparison to the homogeneous bone value. Figure 11b shows the impact of the RL depth on the FAS values. We observe that if we place the RL at a higher cortical layer (grey line), the receivers which first capture the presence of a pore are placed directly above the region of interest showing a velocity decrease of 89 m/s comparing to the reference case of homogeneous bone. However, if we increase the depth from the cortical cortex (purple line), the receivers which are placed at a distance approximately 5 mm from the RL are the first to identify its occurrence showing an increase of the FAS velocity of 51 m/s. Finally, Figure 11c shows that if we change the RL position along the x-axis, the receivers which are placed above the region of interest are the first to capture an increase of the FAS velocity of 23 m/s revealing the position of the larger pore.

Additionally, Figure 11d–f present the local FAS velocity variation profile for the excitation frequency 1 MHz and the presence of a single RL. Specifically, Figure 11d shows that the occurrence of a single RL in the center of the plate (red line) can be identified with a delay by the receivers which follow the region of interest (around x = 23–25 mm) showing an increase of the FAS velocity of 35 m/s. In addition, it can be observed that if we double the diameter of the RL (orange line) the same receivers are the first to identify an increase of the FAS velocity of 65 m/s. In Figure 11e, we observe that if we increase the depth from the cortical surface (purple line) the larger pore cannot be detected as this diagram almost coincides with the case of homogeneous bone. On the contrary, the occurrence of the RL near the upper cortical surface (grey line) leads to a decrease of the FAS velocity of 57 m/s, and it can be identified with a small delay by the receivers placed above the propagation path between x = 21–23 mm. Finally, Figure 11f shows that, if we change the position of the RL along the x-axis, the receivers which are placed between x = 29–31 mm are the first to detect an increase of the FAS velocity of 17 m/s.

Then, in Figure 12, we investigate the potential of the local FAS velocity variation profile to identify the occurrence of a single or a cluster of three or five large pores in the center of the plate taking into account the cortical microstructure. Figure 12a illustrates the local FAS velocities for the excitation frequency 0.5 MHz, while Figure 12b shows the FAS values for 1 MHz considering B_Po5 (red line) as the reference case. For the excitation frequency 0.5 MHz, the relative percentage differences were calculated in the range: (a) 0.05%–0.75% for B_Po5_RL1, (b) 0.05%–1.41% for B_Po5_RL3 and (c) 0.18%–2.71% for B_Po5_RL5. It can be observed that the first receivers which capture the RL are placed between x = 21–23 mm showing a short delay. A 4-mm delay in the detection of the larger pores can be seen in Figure 10b for 1 MHz and lower relative percentage differences were calculated compared to 0.5 MHz from: (a) 0.04–0.32% for B_Po5_RL1, (b) 0.14–0.46% for B_Po5_RL3, (c) 0.05%–0.79% for B_Po5_RL5, respectively.

5. Discussion

This work presents a 2D parametric and systematic computational study aiming to investigate the effect of cortical porosity on ultrasound wave propagation in healthy and osteoporotic long bones. In comparison to previous studies [2,14,26], we established more realistic scenarios, as this is the first time that the distribution of the pores was randomized including normal pores, as well as pores with diameters larger than the Haversian canal size. Moreover, the analysis was not limited to a small segment of cortical bone with the presence of only one large basic multicellular unit as in [14]. The gradual formation of osteoporosis was also considered by simulating more physiological distributions of pores with larger pores in the endosteal cortex and smaller ones in the periosteum. The excitation frequencies 0.5 MHz and 1 MHz were examined to investigate which frequency is more sensitive to detect changes in cortical porosity and the occurrence of RLs. Numerical simulations of wave propagation were performed in the tangential direction and various microstructural models mimicking normal and pathological tissue states were investigated. In comparison to traditional FAS velocity measurements using one source and one receiver, we performed calculations of "local" FAS values in small successive propagation paths along the cortical cortex in order to establish possible guidelines for the interaction of ultrasound with cortical microstructure.

Initially, in Figure 4a,d, and Figure 8a,d, we examined the FAS velocity variation profile along cortical bone for microstructural models with porosities from 0 to 16%. It was made clear that as the porosity increases, the FAS velocities calculated in small propagation paths decrease for both the examined frequencies. However, the relative percentage differences calculated considering B_Hom as the reference case were higher for 0.5 MHz comparing to 1 MHz. Moreover, Figure 7a shows a delay in the time of arrival with increasing the porosity followed by an amplitude increase. This may be explained by the reflection mechanisms, which enhance the received signal for higher porosities. However, despite the significance of the calculation of the FAS velocities from the first signal extremum, the evaluation of the whole waveform would be a possible direction for future research based on guided wave analysis or on the waveform rectification followed by numerical integration via the trapezoidal method (presented in Section 2).

Additionally, the use of implanted transducers led to higher relative percentage differences compared to the use of non-contact transducers, while a delay in the detection of a single or a cluster of RL was observed for non-contact transducers. This can be attributed to the attenuation mechanisms induced by the presence of the soft tissues in the propagation path. The attenuation mechanisms include the evolution of absorption and scattering phenomena. Scattering is induced in bone due to the presence of the pores while soft tissues are major sources of acoustic wave absorption. Despite the accuracy of the FDTD method, which was demonstrated via a convergence study, a limitation of the present work is the simulation weakness of the software to account for attenuation effects resulting from the viscoelasticity [2]. In addition, it should be mentioned that the use of implanted transducers is considered as an invasive procedure for osteoporosis and was examined only for comparison purposes with the non-contact transducers. In clinical practice, their insertion is easy for fracture healing assessment [12,27], but, for the evaluation of osteoporosis, this configuration is not applied during *in vivo* measurements.

Concerning the case of pathology with porosity of 16%, two additional scenarios were investigated to account for the presence of pores with different diameters as well as for the gradual formation of osteoporosis. It was found that the FAS velocity variation profile shows a decrease along the cortical cortex for both B_Po16_RL and B_Po16_Gradual for 0.5 MHz. On the other hand, the excitation frequency of 1 MHz was not convenient for B_Po16_RL (Figure 4e and Figure 8e) as no specific FAS velocity tendency was observed, while, for B_Po16_Gradual, it was sensitive to the gradual porosity increase. A direct comparison of the received signals from B_Po16 and B_Po16_RL was also included inFigure 7b showing a difference in the received signals due to the occurence of large pores.

Additionally, a statistical analysis was conducted based on the estimation of the standard errors, one-way ANOVA and linear regression analysis. It should be mentioned that the receiving positions near the source ($d < 12$ mm) were ignored as in Figure 8a,d, we observe that for short distances between the transducers, low velocities are calculated, revealing that the FAS propagates directly via water. The calculated p-values were lower than 0.05 indicating that there are statistically significant differences between group means as the porosity increases from 0% to 16%. The low p-values can be explained by the high dispersion of the FAS velocity derived from serial measurements for porosity of 16%. Linear regressions analysis revealed a specific FAS velocity trend even for the cases of Figure 4e and Figure 8e. Nevertheless, low R^2 values were derived implying that this analysis could be used only for supplementary observations following traditional local and successive FAS velocity measurements. The establishment of three numerical models for each case is another parameter which requires further discussion. The use of three models is considered as a good compromise to reduce the computational cost and derive accurate conclusions taking into account the calculated standard error values. More specifically, for all the examined cases after three runs, the maximum FAS velocity variation was far less than 5%, thus we decided not to make more runs to keep calculations and simulations simpler.

Then, the case of normal remodeling was examined including the presence of a single or a cluster of RL. Specifically, it was shown that for implanted transducers and 0.5 MHz, as well as for both the hypothetic case of homogeneous and the realistic of nonhomogeneous bone, the presence of one or more RLs can be identified by the receivers which are placed directly above the region of interest revealing the effectiveness of the FAS wave as an indicator. The same observation was made for all the sizes, numbers, depths and lengths of the RL. However, the use of 1 MHz or the transducers' placement into the water led to the detection of larger pores with a short delay when the RLs are placed in the center of the plate (approximately 4 mm) or near the upper cortical cortex (approximately 2 mm). It should be emphasized that the occurrence of a single RL near the lower surface of cortical bone could not be identified when the excitation frequency 1 MHz was applied irrespective of the transducers' configuration. In addition, it was observed that as the receivers are removed from the region of interest the calculated values gradually approximate the values of healthy bone. Nonetheless, the combination of low and high frequencies for the detection of larger pores could be useful, as low frequencies can detect the region of interest along the *x* axis and higher frequencies can reveal the depth of the pores along the *y*-axis. For example, in Figure 9b,e, when the RL is placed near the upper cortical layer (grey line), for both frequencies, we can detect the large pore in the same position implying that the pore is near the cortical upper surface. If we place the RL in the center of the plate (red line), the use of 0.5 MHz frequency leads to the detection of the exact *x* position of the RL, but 1 MHz shows a delay. This reveals that the pore is placed deeper. Finally, Figure 9e shows that for 1 MHz we cannot detect the RL at $y = 3$ (purple line), indicating in combination with 0.5 MHz findings that a large pore exist, but this pore is positioned near the lower surface of the cortical cortex. Thus, even from the limitations of 1 MHz, we may derive specific information concerning the depth of the RL interpreted in combination with 0.5 MHz findings when the occurrence of large basic multicellular units is a factor of major concern.

A possible explanation for the low sensitivity of 1 MHz is that, when the wavelengths are comparable to or smaller ($\lambda_{0.5MHz} = 7.88$ mm, $\lambda_{1MHz} = 3.94$ mm), the thickness of the cortical cortex (4 mm), the FAS wave propagates as a subsurface, lateral wave which travels at the bulk longitudinal velocity of the medium and cannot reflect changes occurring at deeper cortical layers. The calculation of the wavelenths depends on the frequency and the bone's bulk velocity which is 3939 m/s considering the material properties of Table 1. However, bone velocity may vary from 3500 m/s to 4100 m/s depending on the type and anatomic region of bone as well as cortical thickness. According to the literature [15,28], the use of low excitation frequencies (0.2 MHz) has been chosen to yield information on the correlation of the FAS velocity with changes in cortical thickness due to osteoporosis, while it was reported in [28] that the FAS velocities for high frequencies such as 1 MHz were weakly correlated with cortical thicknesses. Therefore, future research should address the question of which frequency range below 0.5 MHz is more convenient for the assessment of osteoporosis considering the occurrence of pores with different sizes.

Concerning the case of homogeneous bone, it was observed that, for 1 MHz, the FAS velocity along the propagation path increases slightly, approximating the bone's bulk velocity (Figure 9d–f), while this is not the case for 0.5 MHz in Figure 9a–c. The observed behavior at 0.5 MHz can be attributed to the fact that the wavelength is almost twice the cortical thickness implying that more complex wave propagation phenomena evolve and the investigation of the propagation of guided waves could convey significant information. On the other hand, for 1 MHz, the FAS propagates as a lateral wave.

Finally, several assumptions were made which require further numerical research and improvements. First of all, although two dimensional geometries were developed, they incorporated various porosity scenarios. Moreover, changes in cortical thickness in cases of pathology in combination with changes in cortical porosity were not considered. Imaging modalities could be also exploited to establish more realistic computational models [29,30]. In addition, the use of point transducers should be mentioned as in experimental procedures the transducers have a finite size. It was reported in [4] that the choice of using point elements instead of finite-sized ones did not affect significantly the recorded ultrasound velocities. According to our previous study using numerical models from SAM images [31], the application of transducers with a finite size larger than the RL diameter could provide significant information concerning the occurrence of a region of RLs. However, further numerical research is needed to investigate the impact of the transducers' size in relation to the scatterers' size on the results. Furthermore, according to [32], the majority of patients with osteoporosis have changes in the mineralization pattern. In particular, reduction of average mineral content was described when compared with agematched controls or when compared with reference values for healthy adults. Quantitative ultrasound was used in [33] for bone characterization and the role of bone mineral density as a major determinant of acoustic properties was highlighted as well as the significance of density-independent relationships with bone microarchitecture. However, according to [34], the main cause of sound attenuation coefficient in bone is porosity, while matrix stiffness has a minor effect. In addition, it was reported in [19] that in aged women the changes in porosity prevail over those of of matrix elasticity to drive the variations of the bone mesoscopic elasticity.

Nevertheless, the results clearly indicated that the excitation frequency 0.5 MHz is more sensitive to detect changes in cortical porosity and the occurrence of RLs. In addition, it was shown that the calculation of the FAS velocity in small propagation paths could potentially provide significant quantitative information for the early diagnosis of osteoporosis and detection of pathologic regions with larger pores with a high risk for fracture.

6. Conclusions

In this work, we presented a systematic and parametric computational study on ultrasound wave propagation in healthy and osteoporotic bones. The FAS velocity was derived in small propagation paths to extract quantitative criteria for the diagnosis and assessment of osteoporosis taking into consideration not only total changes in cortical porosity, but also the occurrence of larger pores and various microstructural features. The role of the excitation central frequency and the impact of the transducers' positions on the FAS velocity variation profile were examined. For the first time, the gradual formation of osteoporosis was also considered. It was shown that the local FAS velocities derived from a proper combination of low and high frequencies may contribute to the prevention or monitoring of osteoporosis and reveal potential pathologic regions with larger pores at an early stage of the disease. Further numerical research is suggested on challenges for clinical application and more realistic scenarios using scanning acoustic microscopy images.

Acknowledgments

This work is part funded by the State Scholarships Foundation and the German Academic Exchange Service and by Deutsche Forschungsgemeinschaft (grant Ra1380/9-1).

Author Contributions

Vassiliki T. Potsika worked on all the sections of the paper. Also, she was mainly responsible for the writing of the manuscript. Konstantinos N. Grivas and Theodoros Gortsas contributed to the development of the numerical models in Matlab and provided the algorithms for the random distribution of the scatterers. Gianluca Iori mainly contributed to Section 2 concerning the benchmark problem using both k-Wave and Simsonic. Vasilios C. Protopappas and Kay Raum set the parameters for the development of innovative numerical models based on realistic scenarios. They provided basic guidelines concerning: (a) the final porosities which were examined describing different stages of the pathology and (b) the selection of the proper numerical tools. Also, they contributed to the presentation of the results and structure of the paper. Demosthenes Polyzos and Dimitrios I. Fotiadis contributed to the assessment of the final results and the evaluation of: (a) the progress of the students and their individual as well as joint projects (b) the writing of the manuscript.

Conflicts of Interest

The authors wish to declare that there is no conflict of interest.

References

1. Laugier, P.; Haïat, G. *Bone Quantitative Ultrasound*; Springer: New York, NY, USA, 2010.

2. Rohde, K.; Rohrbach, D.; Gluer, C.C.; Laugier, P.; Grimal, Q.; Raum, K.; Barkmann, R. Influence of Porosity, Pore Size, and Cortical Thickness on the Propagation of Ultrasonic Waves Guided Through the Femoral Neck Cortex: A Simulation Study. *IEEE Trans. Ultras. Ferroel. Freq. Control***2014**, *61*, 302–313.

3. Grondin, J.; Grimal, Q.; Engelke, K.; Laugier, P. Potential of first arriving signal to assess cortical bone geometry at the hip with QUS: A model based study. *Ultrasound Med. Biol.* **2010**, *36*, 656–666.

4. Moilanen, P.; Talmant, M.; Bousson, V.; Nicholson, P.H.F.; Cheng, S.; Timonen, J.; Laugier, P. Ultrasonically determined thickness of long cortical bones: two-dimensional simulations of *in vitro* experiments. *J. Acoust. Soc. Am.* **2007**, *122*, 1818–1826.

5. Malo, M.K.H.; Rohrbach, D.; Isaksson, H.; Töyräs, J.; Jurvelin, J.S.; Tamminena, I.S.; Kröger, H.; Raum, K. Longitudinal elastic properties and porosity of cortical bone tissue vary

with age in human proximal femur. *Bone* **2013**, *53*, 451–458.

6. Xu, K.; Liu, D.; Ta, D.; Hu, B.; Wang, W. Quantification of guided mode propagation in fractured long bones. *Ultrasonics* **2014**, *54*, 1210–1218.

7. Papacharalampopoulos, A.; Vavva, M.; Protopappas, V.; Polyzos, D.; Fotiadis, D.I. A numerical study on the propagation of Rayleigh and guided waves in cortical bone according to Mindlin's Form II gradient elastic theory. *J. Acoust. Soc. Am.* **2011**, *130*, 1060–1070.

8. Machado, C.B.; Pereira, W.C.D.A.; Granke, M.; Talmant, M.; Padilla, F.; Laugier, P. Experimental and simulation results on the effect of cortical bone mineralization in ultrasound axial transmission measurements: A model for fracture healing ultrasound monitoring. *Bone* **2011**, *48*, 1202–1209.

9. Rohrbach, D.; Lakshmanan, S.; Peyrin, F.; Langer, M.; Gerisch, A.; Grimal, Q.; Laugier, P.; Raum, K. Spatial distribution of tissue level properties in human a femoral cortical bone. *J. Biomech.* **2012**,*45*, 2264–2270.

10. Potsika, V.T.; Vavva, M.G.; Protopappas, V.C. Computational Modeling of Ultrasound Wave Propagation in Bone. In *Computational Medicine in Data Mining and Modeling*; Springer: New York, NY, USA, 2014; pp. 349–376.

11. Moreau, L.; Minonzio, J.G.; Talmant, M.; Laugier, P. Measuring the wavenumber of guided modes in waveguides with linearly varying thickness. *J. Acoust. Soc. Am.* **2014**, *135*, 2614–2624.

12. Protopappas, V.C.; Fotiadis, D.I.; Malizos, K.N. Guided ultrasound wave propagation in intact and healing long bones. *Ultras. Med. Biol.* **2006**, *32*, 693–708.

13. Malo, M.K.H.; Töyräs, J.; Karjalainena, J.P.; Isakssonc, H.; Riekkinena, O.; Jurvelin, J.S. Ultrasound backscatter measurements of intact human proximal femurs—

Relationships of ultrasound parameters with tissue structure and mineral density. *Bone* **2014**, *64*, 240–245.

14. Bourgnon, A.; Sitzer, A.; Chabraborty, A.; Rohde, K.; Varga, P.; Wendlandt, R.; Raum, K. Impact of microscale properties measured by 50-MHz acoustic microscopy on mesoscale elastic and ultimate mechanical cortical bone properties. In Proceedings of IEEE International Ultrasonics Symposium, Chicago, IL, USA, 3–6 September 2014; pp. 636–638.

15. Nicholson, P.; Moilanen, P.; Kärkkäinen, T.; Timonen, J.; Cheng, S. Guided ultrasonic waves in long bones: modelling, experiment and application. *Physiol. Meas.* **2002**, *23*, 755–768.

16. Bossy, E.; Talmant, M.; Laugier, P. Effect of bone cortical thickness on velocity measurements using ultrasonic axial transmission: a 2d simulation study. *J. Acoust. Soc. Am.* **2002**, *112*, 297–307.

17. Moilanen, P.; Talmant, M.; Nicholson, P.H.F.; Cheng, S.; Timonen, J.; Laugier, P. Ultrasonically determined thickness of long cortical bones: Three-dimensional simulations of *in vitro*experiments. *J. Acoust. Soc. Amer.* **2007**, *122*, 2439–2445.

18. Grimal, Q.; Rohrbach, D.; Grondin, J.; Barkmann, R.; Gluer, C.C.; Raum, K.; Laugier, P. Modeling of femoral neck cortical bone for the numerical simulation of ultrasound propagation. *Ultrasound Med. Biol.* **2013**, *40*, 1–12.

19. Granke, M.; Grimal, Q.; Saïed, A.; Nauleau, P.; Peyrin, F.; Laugier, P. Change in porosity is the major determinant of the variation of cortical bone elasticity at the millimeter scale in aged women. *Bone* **2011**, *49*, 1020–1026.

20. Thompson, D.D. Age changes in bone mineralization, cortical thickness, and haversian canal area.*Calcif. Tissue Int.* **1980**, *31*, 5–11.

21. Treeby, B.E.; Jaros, J.; Rendell, A.P.; Cox, B.T. Modeling nonlinear ultrasound propagation in heterogeneous media with power law absorption using a k-space pseudospectral method. *J. Acoust. Soc. Am.* **2012**, *131*, 4324–4336.

22. Bossy, E.; Grimal, Q. Numerical methods for ultrasonic bone characterization. In *Bone Quantitative Ultrasound*; Laugier, P., Haiat, G., Eds.; Springer: New York, NY, USA, 2010; pp. 181–228.

23. Bossy, E.; Laugier, P.; Peyrin, F.; Padilla, F. Attenuation in trabecular bone: A comparison between numerical simulation and experimental results in human femur. *J. Acoust. Soc. Am.* **2007**, *122*, 2469–2475.

24. Muller, M; Moilanen, P.; Bossy, E.; Nicholson, P.; Kilappa, V.; Timonen, J.; Talmant, M.; Cheng, S.; Laugier, P. Comparison of three ultrasonic axial transmission methods for bone assessment.*Ultrasound Med. Biol.* **2005**, *31*, 633–642.

25. Dodd, S.P.; Cunningham, J.L.; Miles, A.W.; Gheduzzi, S.; Humphrey, V.F. Ultrasound transmission loss across transverse and oblique bone fractures: An *in vitro* study. *Ultrasound Med. Biol.* **2008**, *34*, 454–462.

26. Potsika, V.; Spiridon, I.; Protopappas, V.; *et al.* Computational study of the influence of callus porosity on ultrasound propagation in healing bones. In Proceedings of 36th Annual International Conference of the IEEE Engineering in Medicine and Biology Society, Chicago, IL, USA, 3–6 September 2014; pp. 636–638.

27. Potsika, V.T.; Grivas, K.N.; Protopappas, V.C.; Vavva, M.G.; Raum, K.; Rohrbach, D.; Polyzos, D.; Fotiadis, D.I. Application of an effective medium theory for modeling ultrasound wave propagation in healing long bones. *Ultrasonics* **2014**, *54*, 1219–1230.

28. Moilanen, P.; Nicholson, P.H.F.; Kilappa, V.; Cheng, S.; Timonen, J. Assessment of the cortical bone thickness using

ultrasonic guided waves: modelling and *in vitro* study. *Ultrasound Med. Biol.* **2007**,*33*, 254–262.

29. Raum, K.; Grimal, Q.; Varga, P.; Barkmann, R.; Glüer, C.C.; Laugier, P. Ultrasound to Assess Bone Quality. *Curr. Osteoporos. Rep.* **2014**, *12*, 154–162.

30. Raum, K.; Leguerney, I.; Chandelier, F.; Talmant, M.; Saïed, A.; Peyrin, F.; Laugier, P. Site-matched assessment of structural and tissue properties of cortical bone using scanning acoustic microscopy and synchrotron radiation μCT. *Phys. Med. Biol.* **2006**, *51*, 733–746.

31. Potsika, V.T.; Gortsas, T.; Grivas, K.N.; Protopappas, V.C.; Vavva, M.G.; Raum, K.D.; Polyzos, D.; Fotiadis, D.I. Computational modeling of guided ultrasound wave propagation in healthy and osteoporotic bones. In ELEMBIO 6th National Conference, Patras, Greece, 2014.

32. Roschger, P.; Misof, B.; Paschalis, E.; Fratzl, P.; Klaushofer, K. Changes in the Degree of Mineralization with Osteoporosis and its Treatment. *Curr. Osteoporos. Rep.* **2014**, *12*, 338–350.

33. Cortet, B.; Boutry, N.; Dubois, P.; Legroux-Gérot, I.; Cotton, A.; Marchandise, X. Does quantitative ultrasound of bone reflect more bone mineral density than bone microarchitecture. *Calcif. Tissue Int.* **2004**, *74*, 60–67.

34. Sasso, M; Haïat, G.; Yamato, Y.; Naili, S.; Matsukawa, M. Frequency Dependence of Ultrasonic Attenuation in Bovine Cortical Bone: An In Vitro Study. *Ultrasound Med. Biol.* **2007**, *33*, 1933–1942.

© 2016 by the authors; licensee MDPI, Basel, Switzerland. This article is an open access article distributed under the terms and conditions of the Creative Commons by Attribution (CC-BY) license (http://creativecommons.org/licenses/by/4.0/).

11

CHAPTER

The Simple Lamb Wave Analysis to Characterize Concrete Wide Beams by the Practical MASW Test

Young Hak Lee [1] and Taekeun Oh [2,*]

[1]Department of Architecture Engineering, Kyung Hee University, 1732 Deogyeong-daero, Giheung-gu, Yongin 17104, Korea

[2]Department of Safety Engineering, Incheon National University, 12-1 Songdo-dong, Yeonsu-Gu, Incheon 22012, Korea

Abstract

In recent years, the Lamb wave analysis by the multi-channel analysis of surface waves (MASW) for concrete structures has been an effective nondestructive evaluation, such as the condition assessment and dimension identification by the elastic wave velocities and their reflections from boundaries. This study proposes an effective Lamb wave analysis by the practical application of MASW to concrete wide beams in an easy and simple manner in order to identify the dimension and elastic wave velocity (R-wave) for the condition assessment (e.g., the estimation of elastic properties). This is done by identifying the zero-order antisymmetric (A0) and first-order symmetric (S1) modes among multimodal Lamb waves. The MASW data were collected on eight concrete wide beams and compared to the actual depth and to the pressure (P-) wave velocities collected for the same specimen. Information is extracted from multimodal Lamb wave dispersion curves to obtain the elastic stiffness parameters and the thickness of the concrete structures. Due to the simple and cost-effective procedure associated with the MASW processing technique, the characteristics of several fundamental modes in the experimental Lamb wave dispersion curves could be measured. Available reference data are in good agreement with the parameters that were determined by our analysis scheme.

Keywords: Lamb wave; MASW; surface wave; Rayleigh wave; P-wave; condition assessment; concrete; thickness estimation

1. Introduction

Concrete has been a popular construction material due to its economic efficiency and good durability. As such, concerns about concrete structures, with regard to their quality assurance (in terms of the strength, elasticity and density) and their condition assessment (in terms of dimension identification and the damage characterization of delamination, cracks and voids), are important to consider [1,2]. Various non-destructive evaluation (NDE) methods, as a solution to these types of issues, have been proposed and developed [3]. Among these, stress wave-based testing techniques have the advantage of allowing one to determine certain fundamental material properties and dimensions, as well as identify flaws, in a simple and effective manner [4]. As an example of a stress wave-based NDE, the impact-echo (IE) method, which uses multiple body wave reflections from the crack or bottom surface, has been widely applied to concrete structures in order to determine dimensions and to characterize damage. In the IE measurement, the tested structure is subjected to a transient impact. The subsequent transient dynamic response is recorded, and the resonance frequency of the structure is determined. The resonance frequency corresponding to the impact-echo mode (or the thickness mode), and the indirectly-measured P-wave velocity can indicate certain material properties, as well as the thickness or existence of flaws [5]. Alternatively, the ultrasonic testing method, which utilizes the velocity and reflections of pressure, shear and Rayleigh waves (P-, S- and R-waves), can provide useful information about the dimensions and structural integrity. Initially, the P-wave was commonly applied to concrete structures, but the application of S- or R-waves has gradually increased because these are less sensitive to confinement and boundary conditions; they also have a larger energy magnitude than P-waves [6].

For plate-like structures, such as concrete slabs and asphalt pavements, the IE and ultrasonic testing methods can be effectively applied, and the combination of these two techniques can give more accurate information. In this regard, many researchers have studied the fusion of various NDE methods. Kee *et al.* [7] proposed combining impact-echo and infrared thermography for damage identification in the deck slabs of bridges. Baggens and Ryden [8] presented the possibility of a combined impact-echo and surface wave measurement to analyze plate-like structures, such as asphalt pavement, and to identify Lamb wave plate parameters, such as Poisson's ratio, because the asphalt has a broad range of Poisson's ratio due to the high sensitivity of temperature. They utilized both the Rayleigh wave velocity from A0 Lamb mode by the MASW test and the ratio of S1-ZGV (zero group velocity) Lamb mode frequency to the A2-ZGV one by the impact-echo test for the estimation of Poisson's ratio. These efforts have contributed greatly to the development of NDE methods for plate-like structures; however, they still have some limitations in practical field applications, including their complicated interpretation, laborious testing and indirect information.

This study focuses on the simple Lamb wave analysis by the practical application of MASW, combining the IE method and the Rayleigh wave velocity measurement, which utilizes simple equipment and a facile analysis process without the separate impact-echo test. The proposed MASW technique yields effective phase dispersion curves not only of the plate-like one, such as the slab and pavement, but also of the semi-plate-like structures, such as the wide-beam, by minimizing the effects of reflections from edge boundaries through the coherence of many data acquisitions associated with the multiple sensors and instrumented impact hammer with a high frequency range in a test setup. Among various guided Lamb modes, the A0 and S1 modes, which correspond to the Rayleigh wave velocity and impact-echo mode, respectively, were easily and clearly identified. In a field test, the

A0 and S1 modes extracted from dispersion curves (made from field measurements) can be used to characterize a material or provide quality assurance. Moreover, this information can be used to make calibration values for the elastic properties and thickness, which is useful in conventional ultrasonic testing, especially if the thickness of a structure is not known in advance. Furthermore, the entire testing procedure can be completed in an hour with cost-effective equipment. Both measurements and analysis are made on the same portable computer, and the recorded data can be evaluated directly in the field to obtain the final results. The experimental results present practical guidelines for applying stress wave-based NDE methods to concrete structures.

2. Lamb Wave Modes and Dispersion Curves

The dispersion curve obtained by the MASW in this study is based on the principles of Lamb wave theory. Lamb waves [9] are a type of guided dispersive waves propagating in elastic and isotropic plate structures with traction-free boundaries. Lamb waves formed by the interference of multiple reflections and the mode conversion of P- and S-waves at the free surfaces of the plate are presented in Figure 1[10]. Typically, by matching theoretical multimodes of Lamb wave dispersion curves with experimental ones, the shear (V_S) and Raleigh wave (V_R) velocities, Poisson's ratio (v) and the thickness of the tested plate structure can be determined. Lamb waves have been used in various NDT applications, such as material characterization of elastic plates [11,12], defect inspection [13] and thickness measurements of thin films [14].

Lamb waves are a group of wave types including the bending wave, the Rayleigh wave and the quasi-longitudinal wave. Lamb [9] organized the dispersion equation that manages the transitions between these types of waves. Harmonic wave propagation in the x-direction can be represented as combinations of the frequency (f) and phase velocity (C_P) that correspond to standing waves in the thickness direction (y-direction). These waves must obey the following dispersion Equations [9]:

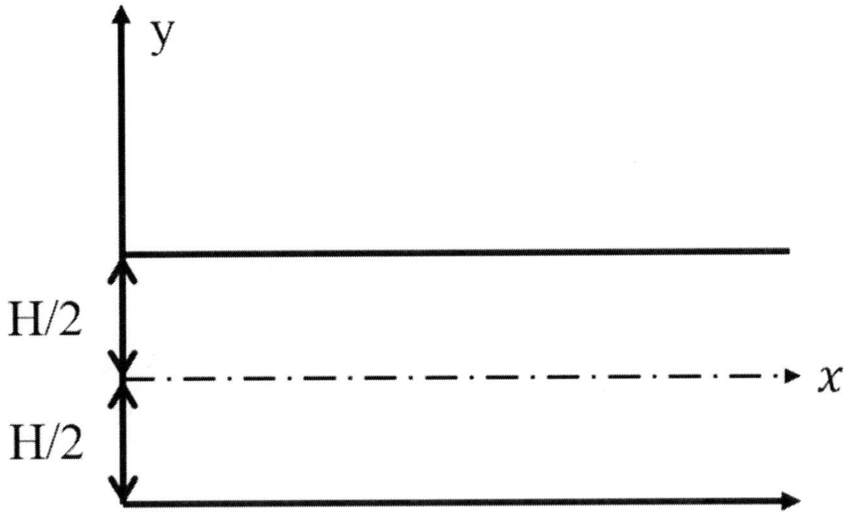

Figure 1. Schematic geometry of a free plate problem with regard to the Lamb wave.

$$\frac{\tan\left(q\frac{H}{2}\right)}{\tan\left(p\frac{H}{2}\right)} = -\frac{4k^2pq}{\left(q^2-k^2\right)^2} \text{ for symmetric modes} \quad (1)$$

$$\frac{\tan\left(q\frac{H}{2}\right)}{\tan\left(p\frac{H}{2}\right)} = -\frac{\left(q^2-k^2\right)^2}{4k^2pq} \text{ for symmetric modes} \quad (2)$$

Here, p and q are given by:

$$p^2 = \left(\frac{\omega}{V_P}\right)^2 - k^2 \quad (3)$$

$$q^2 = \left(\frac{\omega}{V_S}\right)^2 - k^2 \quad (4)$$

The wavenumber k is numerically equal to ω/C_P, where C_P is the phase velocity of the Lamb wave mode and ω is the circular frequency. The phase velocity is related to the wavelength by $C_P = (\omega/2\pi)/\lambda$. V_P and V_S are the pressure and shear wave velocities, respectively.

The dispersion relation is a transcendental function; it is not straightforward to calculate a k value at any given frequency. A root-searching technique must be used to find the correct wave numbers at any given frequency [15]. By extracting roots over a wide frequency range, dispersion curves of both symmetric (S) and antisymmetric (A) Lamb waves can be studied. Figure 2 shows the mode shapes of symmetric (S0) and antisymmetric (A0) fundamental Lamb wave modes. Longitudinal displacements of symmetric and antisymmetric wave modes are equal and opposite, respectively, on either side of the median plane. Alternatively, transverse displacements are opposite and equal, respectively [16].

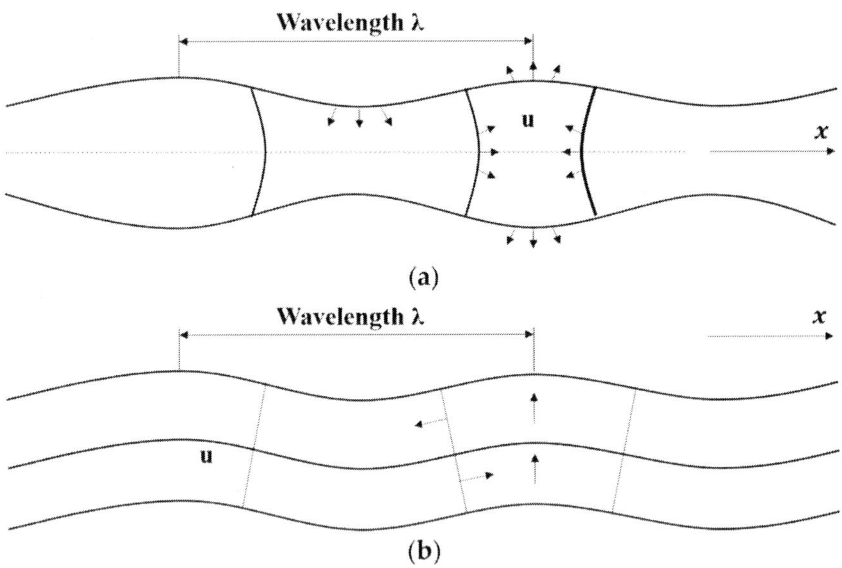

Figure 2. Mode shapes of symmetric (S0) and antisymmetric (A0) fundamental Lamb wave modes. (**a**) Symmetric mode; (**b**) antisymmetric mode.

Figure 3 presents the dispersion curve of a concrete slab with elastic properties corresponding to normal-strength concrete. At the lowest frequencies (<5 kHz), only the two fundamental modes A0 and S0 exist. In this frequency range, the S0 dispersion curve approaches the quasi-longitudinal wave velocity, which is the P-wave velocity along the plate. The A0 dispersion curve approaches the bending wave velocity of the plate: 300 m/s to 400 m/s at these low frequencies. At higher frequencies, A0 and S0 approach the Rayleigh wave velocity of the plate [17]. It can be shown analytically that Equations (1) and (2) reduce to the Rayleigh wave dispersion equation in a homogeneous half space when the frequency approaches infinity. Typically, in this example, Rayleigh wave motion develops at a range of 10 to 12 kHz; this is where A0 and S0 merge together at the Rayleigh wave velocity. Gibson and Popovics [18] showed that the impact-echo resonance actually corresponds to the S1 mode Lamb wave at the zero group velocity (ZGV) frequency condition. Thus, the S1 -ZGV mode frequency can be used to estimate the plate thickness. In Figure 3, the impact-echo frequency corresponds to 7 kHz in the S1 mode, and the thickness is calculated to be 0.3 m by using the impact-echo equation (Equation (5)). This indicates that the slab thickness H is related to the P-wave velocity V_P and the impact-echo frequency f_{IE}.

$$H = \frac{\beta V_P}{2 f_{IE}} \tag{5}$$

For plate-like structures, β is approximately 0.96 [5].

At these high frequencies and small wavelengths, the finite-thickness plate appears as a semi-infinite medium, and wave propagation is limited to the regions closest to the surfaces. Higher modes develop at their respective cut-off frequencies, which are related to the thickness of the plate [19]. At higher frequencies, all higher modes approach the shear wave velocity of the plate. It should also be noted that all symmetric modes have an almost straight portion where the phase velocity is close to the quasi-longitudinal wave velocity.

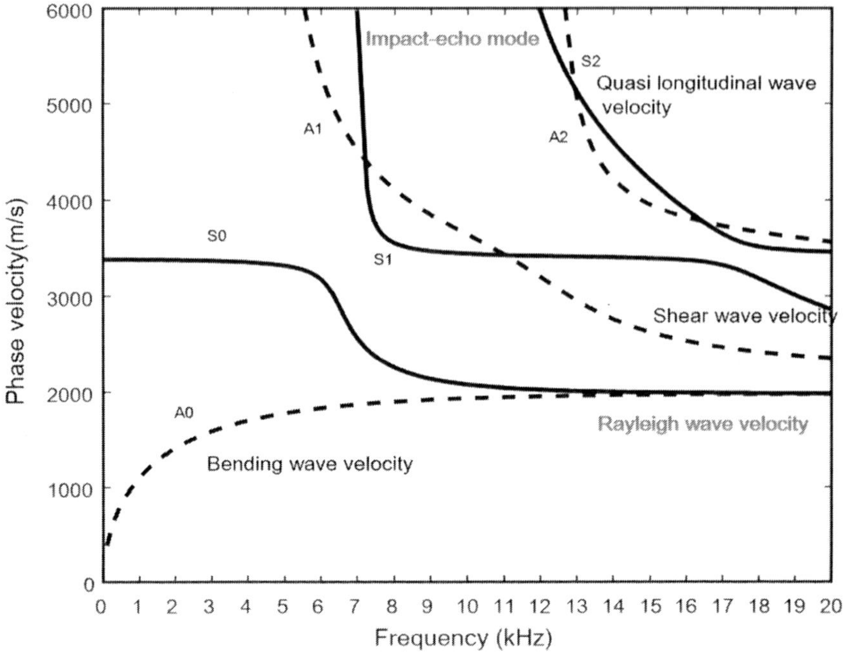

Figure 3. Phase velocity dispersion curves determined by MASW (based on the FE simulation of a concrete slab with a thickness of 0.3 m). A density of 2250 kg/m^3, an elasticity of 23 GPa and a fixed Poisson's ratio of 0.167 are assumed for the concrete slab.

3. Measurement of the R-Wave Velocity and Thickness by MASW

The conventional MASW measurement for the dispersion curve of Lamb waves in the plate is based on N-wave signals collected on the surface along a linear array of sensed points that are equally spaced (with spacing dx) from the wave source [20]. Multiple data are processed as individual signals obtained by each sensor. Data are transformed from the offset time domain to the frequency phase velocity domain using a 2D Fourier transform, which produces a phase velocity that is typically referred to as the dispersion curve [21].

In this study, an effective testing procedure that is associated with the frame equipped with all of the sensors was applied to the

MASW measurement. Using this technique, improved dispersion curves could be measured; this is referred to as "multichannel data acquisition with a frame system (MDAFS)". The multimodal nature of Lamb wave propagation implies that a transient source (e.g., an impact hammer) at the surface will easily generate multiple modes of propagation. Measured time histories along the surface will be composed of a number of superposed modes. Therefore, it is important that different modes of Lamb wave propagation be fulfilled with a practical and cost-effective approach.

The scheme of MDAFS is presented in Figure 4. This makes it possible for a multichannel record to be obtained with a single test. All of the recorded signals are compiled to make an equivalent multichannel record for dispersion analysis. With this system, the signal from each impact is automatically streamed to the hard drive of a portable computer. The resulting multichannel record at different offsets, $u(x,t)$, is automatically and objectively transformed to the frequency phase velocity domain with the MASW processing technique; the plane wave transformation is in the following equation [20,21].

$$S(\omega, c_P) = \int e^{-i(\omega/c_P)x} U(x,\omega)dx \tag{6}$$

where $U(x,\omega)$ is the normalized complex spectrum obtained from the Fourier transformation of $u(x,t)$,ω is the angular frequency and C_P is the phase velocity.

$S(\omega,C_P)$ is the stack amplitude for each ω and C_P, which can be viewed as the coherency in the linear arrival pattern along the offset range for that specific combination of ω and C_P. Thus, it is possible to extract multimodal dispersion curves from a multichannel dataset. All coherent phase velocity patterns are mapped with respect to their relative energy level. Therefore, the phase velocity image shows how the total seismic energy is distributed between different frequencies and phase velocities.

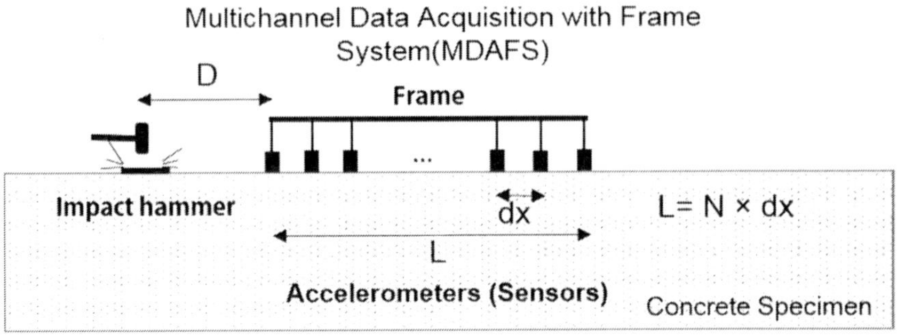

Figure 4. The scheme for an MASW test associated with MDAFS. D is near the offset; dx is the sensor spacing; and L is the offset range.

As a next step, the measured Lamb wave dispersion curves are matched to theoretical dispersion curves. In the general inversion analysis of MASW for the characterization of soil profile and asphalt pavement, by changing the density, elasticity, thickness and Poisson's ratio, the theoretical dispersion curves are matched (either manually or automatically) to the experimental phase velocity image [20,21]. However, this study utilizes the measured Poisson's ratio by the standard test or the representative value (=0.2) for the normal-strength concrete, because the change of the dispersion curve is negligible in the general range of Poisson's ratio 0.15~0.25 for the concrete [17]. Moreover, Popovics [22] showed that both the static and dynamic Poisson's ratio vary across a broad range of values for concrete, usually between 0.15 and 0.25, and that the dynamic and static values of Poisson's ratio are not strongly related to each other.

Lastly, the elasticity can be estimated by Poisson's ratio and the Rayleigh wave velocity extracted from the A0 Lamb mode in the high frequency range; then, the thickness of the target structure can be estimated by Equation (5), combing the estimation of P-wave velocity from R-wave velocity and the S1-ZGV Lamb mode frequency. This whole procedure can be done in the field in just a few minutes, provided that normalized dispersion curves have been calculated in advance (as described in the previous section).

Thus, the Rayleigh wave velocity and thickness can be estimated from the A0 and S1 modes in the dispersion curve. The complete analysis scheme is discussed below.

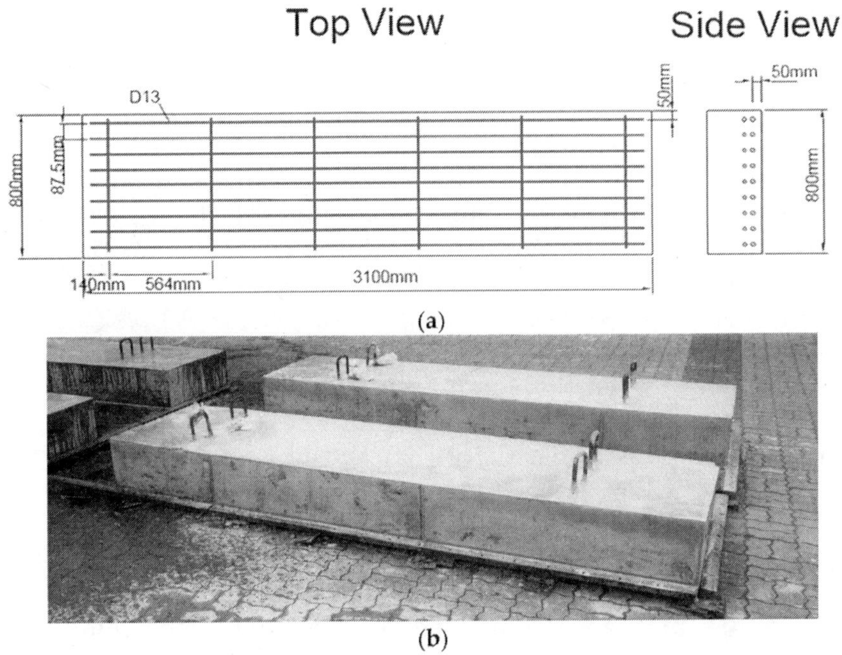

Figure 5. Concrete wide beam test specimen. (**a**) The planar and side views; (**b**) the test specimen after casting.

4. Experiments

4.1. Test Specimens

To investigate the practical effectiveness of our MASW scheme for assessing the conditions of concrete structures, we carried out a series of MASW tests on eight concrete wide beams (W1, W2, ..., W8) cast from the same batch of concrete, as shown in Figure 5. All of the specimens are 800 mm × 3100 mm with a nominal thickness of 300 mm. They all contain transverse and longitudinal 13 mm-diameter reinforcements with 564 mm and 87.5 mm spacings, respectively, at a depth of 250 mm.

The specified w/c of the concrete mixture was 0.70 with an entrained air content of 6%. The mixture contained Type I cement (Ssangyong Cement Co., Ltd., Yeongwol, Korea) and limestone aggregates with a maximum coarse aggregate size of 25 mm. At the time of concrete placement, the slump was measured to be 150 mm, and the density was 2250 kg/m^3. The specimens were moist-cured for seven days by covering them with saturated burlap with a bleed hose and plastic sheeting. Beyond the eighth day, the specimens were kept in laboratory air.

We determined the mechanical properties of the concrete by conducting tests performed on ten 150 mm × 300 mm standard cylinders (Kumhohoning, Seoul, Korea) prepared from the same batch of concrete. For the same conditions as the W1, W2, ..., W8 concrete wide beams, we air-cured the standard specimens for 28 days. We determined the compressive strength using ASTM C-39 (ASTM International, West Conshohocken, PA, USA) [23] and found the modulus of elasticity (Ec) and Poisson's ratio (v) with ASTM C-469 (ASTM International, West Conshohocken, PA, USA) [24]. The average results for the 28-day tests yielded 14.84 MPa for the strength, 17.68 GPa for the modulus of elasticity and 0.167 for static Poisson's ratio. Furthermore, the dynamic modulus (Ed) was 21.30 GPa, as determined by following the empirical relationship proposed by Lydon and Balendran [25].

$$E_c = 0.83\, E_d\ (\mathrm{GPa}) \tag{7}$$

4.2. Test Methods

The purpose of this study is to verify the practical effectiveness of our proposed MASW in the easy and simple manner associated with the MDAFS. First, the condition assessment and dimension identification of concrete structures were carried out by MASW. Second, the P- and S-wave velocities were also measured to verify the test results.

4.2.1. MASW Test for R-Wave Velocity and Thickness Measurement

We applied an MDAFS testing configuration to consistently identify the R-wave velocity and thickness of eight test specimens along four different paths (I, II, III and IV), as shown in Figure 6.

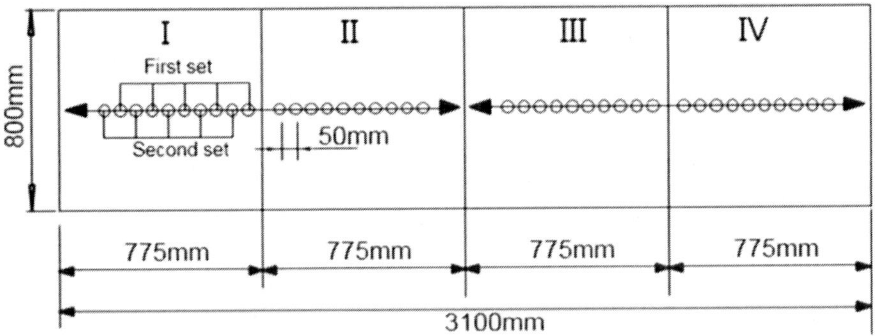

Figure 6. A planar view of the four test paths for the MASW test. The O marks indicate the sensor positions in each path, and the arrows show the direction of each test.

We used a ball-type impact hammer, which is capable of measuring the force historical data for each MASW test. The forcing function associated with the impact exhibits consistent and broad spectral content, ranging from Direct Current (DC) to 30 kHz. We used an accelerometer (PCB 353B16) (PCB Piezotronics, Inc., Depew, NY, USA) with a ±5% frequency range of 1 kHz to 60 kHz and a resonance frequency around 70 kHz. Due to the number of sensors, five accelerometers with 10 cm spacing were equipped with the MDAFS. A dry point coupling between the accelerometer and concrete surface was done by both the thin rubber tip on each sensor and spring loading from the frame. For improved application of MASW, the MDAFS was carried out twice with a 5 cm offset along the same path. In this manner, 10 sensor measurements with 5 cm spacing along the test line could be completed. Figure 7shows the MDAFS testing configuration. As a next step, we stabilized the signals using a signal conditioner (PCB 482C16) (PCB Piezotronics, Inc., Depew, NY, USA) and digitized them at a sampling frequency of 1 MHz using an NI-PXI 5105 oscilloscope (National Instruments Co., Austin, TX, USA). The basic concept in our MASW setup is similar to the classical procedure, which consists of one receiver (geophone) and the multiple hammer impacts in the geotechnical engineering [20,21]. However, the frequency range of the sensor, impactor and Data Acquisition (DAQ) in our test is tailored to the concrete structures with normal or high strength. Furthermore, the simultaneous data record by the multiple sensors associated with the high sampling rate makes it possible to have a clear experimental dispersion curve in the high frequency range.

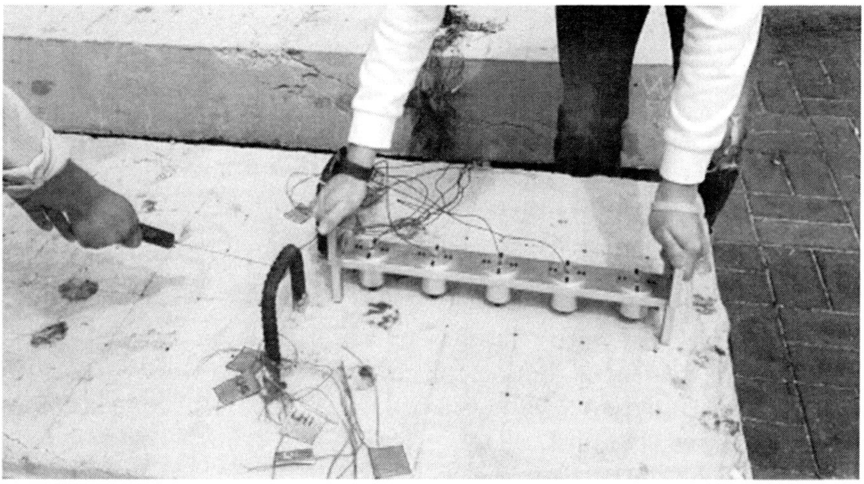

Figure 7. The MDAFS testing configuration for the MASW along predefined paths on the surface of a concrete wide beam.

The multichannel record consists of 10 time series (called traces) from the receivers in an ordered manner. The overall MASW data processing and analysis consists of four steps: (1) the multichannel data record; (2) construction of the dispersion image and extraction of the signal dispersion curve; (3) identification of several fundamental Lamb wave modes; and (4) estimation of the R-wave velocity and impact-echo frequency from A0 and S1 modes. All of these steps can be fully automated. Components of the necessary dispersion curve, such as the A0, A1, S0 and S1 modes, are then extracted from the accumulation pattern. The extracted dispersion curve is finally used as a reference to identify the R-wave velocity and thickness, corresponding to A0 and S1, respectively. For each test, we obtained time signals that were 10 ms in duration with a 1-MHz sampling rate. Because the amplitude of the input forcing function at each test point across the concrete surface is inherently inconsistent, we normalized the amplitude of each time signal with respect to the maximum peak of the input force from the force data measured by the impact hammer; this provides more consistent MASW data.

Figure 8 shows the experimental MASW results along Path 3 (III) in the W2 beam and exhibits dispersion images and Lamb wave multimodes up to 20 kHz. The Lamb wave curves in the single layer must be fit to one of the responses. The A0 and S0 modes converge beyond 10 kHz into a phase velocity of 1878 m/s, which corresponds to the R-wave. The P-wave velocity (3279 m/s) can be estimated from the R-wave velocity by using the elastic relationship described below Equation (8) (with Poisson's ratio equal to 0.167, which was already obtained from ASTM C-469 (ASTM International, West Conshohocken, PA, USA). Furthermore, Poisson's ratio can be verified by matching the experimental dispersion curves with the theoretical Lamb wave modes, which is done by changing the elastic properties (e.g., the elasticity, density and Poisson's ratio). This process is typically referred to as inversion analysis.

$$V_P = \frac{1+\nu}{0.87+1.12\nu}\sqrt{\frac{2(1-\nu)}{1-2\nu}}V_R \tag{8}$$

To estimate the thickness, the S1-ZGV Lamb mode frequency that corresponds to the impact-echo mode (thickness mode) is identified to be 5.1 kHz (Figure 8). Using the impact-echo Equation (5), the thickness is then calculated to be 0.309 m.

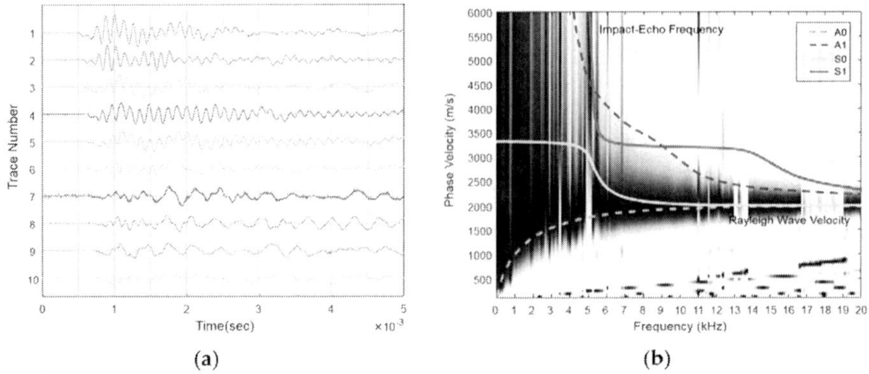

Figure 8. Experimental MASW results along Path III in specimen W2. The computed A0, A1, S0 and S1 Lamb modes corresponding to the specimen depth overlap on the dispersion curve image. (**a**) The signals in the time domain; (**b**) dispersion curve.

4.2.2. P-Wave Velocity Measurement

For reference, the P-wave velocity of concrete V_P (according to British Standard (BS) 1881) [26] was measured using a pair of P-wave transducers (MK-954 transmitters and receivers) (MKC Co., Seoul, Korea) that were connected to a pulse-receiver (Ultracon-170) (MKC Co., Seoul, Korea). The transmitter was driven by a 200-V square pulse, generating a transverse ultrasonic pulse of 52 kHz. The receiver measured the transient stress waves through the surface of each concrete beam (Figure 9).

We performed P-wave measurements with a digital scope board (National International Co., Austin, TX, USA) synchronized with the pulser-receiver unit; therefore, data collection started at the time of pulse application. Measurement of the flight time was affected by electrical noise superimposed with the waveform. Therefore, to minimize random errors associated with identifying the arrival time, we performed wave-averaging on 128 waveforms. The first step in identifying the flight time from the waveform was to establish a baseline. Afterward, the waveform was further smoothed using a 10-point moving average filter.

Figure 9. P-wave measurement in a grid on the surface of a concrete wide beam.

We made a total of 24 indirect measurements on each concrete wide beam using a coordinate system drawn on the beam's surface

(Figure 10). The coordinate system consisted of a primary grid with 100 × 113 mm² spacing. The gridlines were labeled along the width of the specimens as Axes A, B, C, D, E and F and along the length of the specimens as Axes I, II, III and IV. We made indirect measurements longitudinally along the lettered axes. We placed the transmitter and receiver transducers at the grid nodes and measured the average P-wave velocity between them.

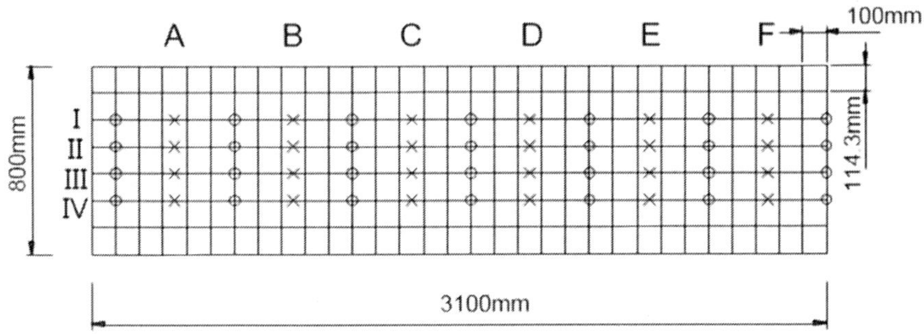

Figure 10. A planar view of the 114 × 100 mm² test point grid for P-wave measurements. The test was carried out on the grid locations. The O and X marks indicate the positions of the transducers and average measurements, respectively. All dimensions in millimeters.

We acquired the transmitter and conditioned receiver signals using a high-speed (1-MHz sampling rate) analog-to-digital data acquisition board (National International Co., Austin, TX, USA). We developed a computer algorithm based on a fixed threshold level to determine the time of flight using the digitally-acquired waveforms.

5. Results and Discussion

5.1. Experimental Results from the MASW Tests

Generally, a variety of experimental data along each test line from the MASW test is presented in the dispersion curves of various

Lamb wave modes. Then, the data are fit to analytically-computed fundamental A0, A1, S0 and S2 Lamb modes with an iterative procedure by changing the Poisson's ratio, density and elasticity of the material. However, in this study, a much simpler procedure than the general approach was applied. This consists of four steps: (1) identification of the Rayleigh wave velocity, which is the asymptotic value of the phase velocity of the A0 mode at increasing frequencies in the dispersion curve; (2) estimation of the P-wave velocity with the elastic wave relationship associated with the Poisson's ratio by Equation (8); (3) identification of the S1-ZGV Lamb mode frequency; and (4) estimation of the thickness with the impact-echo Equation (5). Although the measurement of Poisson's ratio is not carried out, the average value in the range of 0.15 and 0.25 for the normal strength concrete can be used because the dispersion curves are insensitive to changes in Poisson's ratio [17]. In addition, the ZGV frequency of the S1 mode can be reconfirmed by relating the maxima associated with the fundamental thickness resonance in the summed frequency domain amplitude spectra. Thus, the proposed analysis technique is straightforward and can be easily applied.

Figure 11 shows the test results from two representative samples obtained by the proposed procedure. This figure easily confirms that the asymptotic phase velocity over 10 kHz approaches the Rayleigh wave velocity (near 1900 m/s) and that the impact-echo frequency (near 5 kHz) in the frequency domain amplitude spectra is well-matched with the ZGV frequency of the S1 mode in the dispersion curve. In addition, the cut-off region (white area between 5 and 6 kHz) in the A0 mode also guarantees the existence of the ZGV frequency in the S1 mode. The amplitude at the impact-echo frequency in the frequency domain amplitude spectra along Path III in the W2 beam indicates one representative thickness, whereas the amplitude along Path II in the W4 beam is considerably lower; this is the case because the thickness cannot be consistent along Path III. Utilizing the impact-echo Equation (5) with the impact-echo frequency and the P-wave velocity estimated from the R-wave velocity, the thickness is estimated to be about 0.3 m; this is close to the actual thickness.

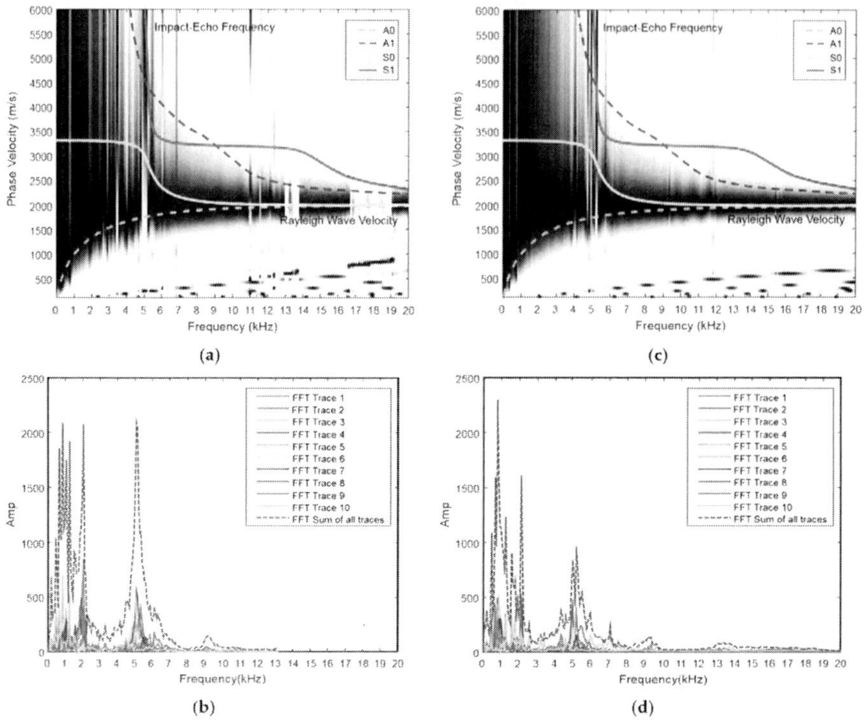

Figure 11. Experimental MASW results for two concrete wide beams (W2 and W4). A Poisson's ratio of 0.16 was measured for the concrete cylinder. (**a**) Phase velocity dispersion curve along path III in the W2 beam; (**b**) summed amplitude spectra along path III in the W2 beam; (**c**) phase velocity dispersion curve along path II in the W4 beam: (**d**) summed amplitude spectra along path II in the W4 beam.

Figure 12 shows the change of the theoretical dispersion curve for the three values of Poisson's ratio (0.16, 0.18 and 0.20) on the experimental dispersion spectrum along Path III in the W2 beam. The phase velocities (Rayleigh wave velocity) of the A0 Lamb mode at the 20 kHz for the Poisson's ratios 0.16 and 0.20 are 1878 m/s and 1860 m/s, and S1-ZGV Lamb wave frequencies for all of the Poisson's ratios are converge to 5100 kHz. Thus, the dispersion curve is not sensitive to a general range of Poisson's ratio for the concrete.

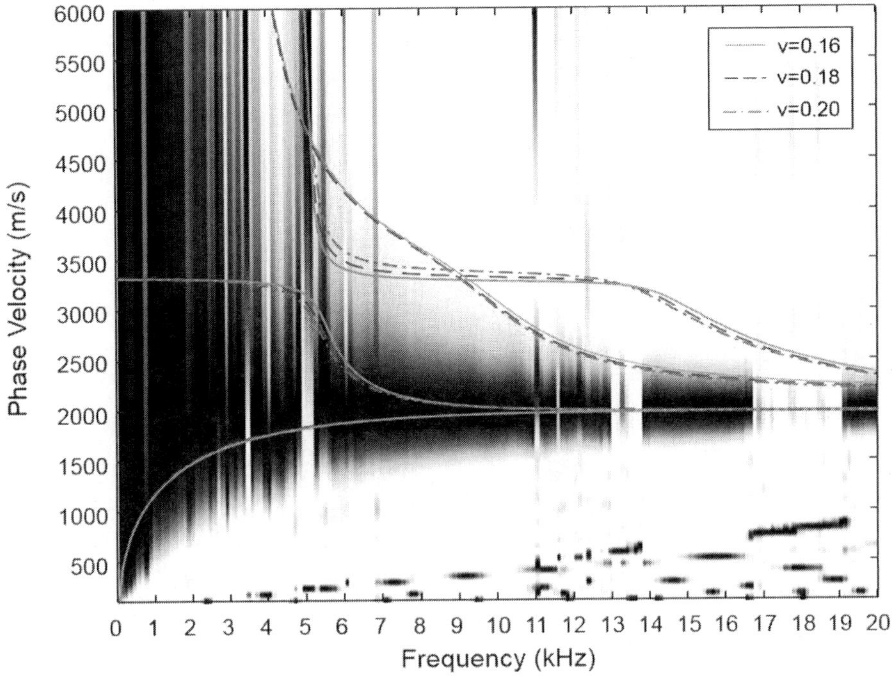

Figure 12. The theoretical dispersion curves on the experimental dispersion spectrum for three values of Poisson's ratios in the W2 concrete wide beam.

The data from the dispersion curve and the summed amplitude spectra below 3 kHz show lots of noise and several large peaks that are not related to our interests (*i.e.*, the Rayleigh wave velocity and ZGV frequency of the S1 mode). These could be caused by the bending modes of the beam itself (flexural behavior) and/or by interference between the Rayleigh wave and the P-wave close to impact, which is typically considered to be a near-field effect. Fortunately, separating this noise from the impact-echo frequency (5 kHz) was unnecessary in our specimens. However, for the structures for which the width to depth ratio is small (*i.e.*, far from the plate-like structure), signal processing (e.g., filtering) can be necessary. Thus, the MASW test has certain limitations in applying to no plate-like structures. Table 1shows a summary of the MASW results for all of the specimens (W1 to W8). Due to the advantage of multiple sensors, the measured Rayleigh wave velocities and thicknesses are consistent.

Table 1. The experimental results of MASW along four paths in all of the specimens. VR, Raleigh wave velocity.

Specimen	**W1**				**W2**				**W3**				**W4**			
Test point	I	II	III	IV	I	II	III	IV	I	II	III	IV	I	II	III	IV
VR (m/s)	1842	1842	1829	1899	1828	1873	1878	1810	1840	1892	1881	1842	1859	1874	1833	1862
Avg.* VR	1853				1847				1864				1857			
Vp** (m/s)	3216	3216	3194	3316	3192	3271	3279	3161	3213	3304	3284	3216	3246	3272	3201	3251
f_{IE} (Hz)	5100	5100	5300	5600	5700	5300	5100	5500	5500	5100	5400	5000	5000	5600	5100	5500
Depth (cm)	0.303	0.303	0.289	0.284	0.269	0.296	0.309	0.279	0.280	0.311	0.292	0.309	0.312	0.280	0.301	0.284
Avg.* thickness	0.295				0.287				0.298				0.294			
Specimen	**W5**				**W6**				**W7**				**W8**			
Test point	I	II	III	IV	I	II	III	IV	I	II	III	IV	I	II	III	IV
VR (m/s)	1866	1905	1845	1880	1884	1858	1874	1879	1869	1888	1880	1832	1885	1862	1856	1885
Avg.* VR	1874				1874				1867				1872			
Vp** (m/s)	3258	3326	3222	3283	3290	3244	3272	3281	3264	3297	3283	3199	3291	3251	3241	3291
f_{IE} (Hz)	5500	5600	5400	5000	5100	5000	5100	5500	5000	5200	5600	5700	5500	5600	5400	5000
Depth (cm)	0.284	0.285	0.286	0.315	0.310	0.311	0.308	0.286	0.313	0.304	0.281	0.269	0.287	0.279	0.288	0.316
Avg.* thickness	0.293				0.304				0.292				0.292			

Avg.* = Average, Vp** = P-wave velocity estimated by Equation (8).

5.2. Experimental Data of P-Wave and the Statistical Reliability

To verify the MASW results, the P-wave velocities were also measured. The data reliability in the measurement of the P-wave velocity were assured by calculating the averaged flying time based on 128 waveforms. The range of the P-wave velocities for all of the specimens is 3263 m/s to 3471 m/s. This range is comparable to the typical one of sound concrete [27]. Furthermore, we investigated the statistical reliability analysis for all of the Ultrasonic Pulse Velocity (UPV) data, including the R-wave velocities (by MASW) and the P-wave velocities. This was done in order to evaluate variations in the experimental data. We used the sample size, mean and standard deviation of the UPV for the basic inputs, as summarized in Table 2.

Table 2. Results of the statistical analysis for R-wave velocities.

Specimen		W1	W2	W3	W4	W5	W6	W7	W8	Total Avg.
R-wave by MASW	Sample size	4	4	4	4	4	4	4	4	4
	Avg. (μ) * (m/s)	1853	1847	1864	1858	1874	1874	1867	1872	1864
	Std (σ) ** (m/s)	31	34	27	19	25	11	25	15	23
	COV	1.7%	1.8%	1.4%	1.0%	1.3%	0.6%	1.3%	0.8%	1.3%
P-wave by UPV	Sample size	24	24	24	24	24	24	24	24	24
	Avg. (μ) * (m/s)	3471	3359	3275	3327	3272	3263	3274	3341	3323
	Std (σ) ** (m/s)	232	231	207	302	157	153	176	122	197
	COV ***	6.7%	6.9%	6.3%	9.1%	4.8%	4.7%	5.4%	3.7%	5.9%

Avg. (μ) * = average velocity; Std (σ) ** = standard deviation of velocity; COV *** = coefficient of the variable ($=\sigma/\mu \times 100$ (%)).

The statistical data show that the standard deviation and COV decrease in the order of R- and P-waves, which means that the UPV data are more statistically stable in the order. According to American Concrete Institute (ACI) 214R, the COVs of the R-wave velocities are excellent (<3.0), and the COVs of the P-wave velocities are fair (5.0 to 6.0) [28]. In summary, based on the respective means and standard deviations, we infer that all of the UPV data are statistically reliable.

5.3. Comparison with the P-Wave Velocities and the Actual Depth

We checked the accuracy of our experimental MASW data by comparing the R-wave velocity and thickness with the measured P-wave velocities (obtained from traditional UPV measurements) and the actual thickness. Table 3 shows the comparison between P-wave velocities (estimated from the R-wave velocity) and the measured P-wave ones. It also compares the estimates from the MASW data to the actual thickness. The average error of the comparisons with the measured P-wave velocities is 2.09%. Furthermore, the error with regard to the thickness estimate is only 2.17%. Thus, the results from the MASW test are comparable to the results of other UPV tests. The practical application of MASW that is proposed in this study can be an effective method to identify the dimensions and elastic wave velocities.

Table 3. Comparison of MASW results with the measured P-wave velocities and the actual thickness.

Specimen	MASW			Vp (m/s)	Actual Thickness (m)	Error (%)	
	VR (m/s)	Vp * (m/s)	Thickness (m)			P-Wave	Thickness
W1	1853	3236	0.295	3471	0.3	6.77%	1.76%
W2	1847	3226	0.287	3359	0.3	3.98%	4.21%
W3	1864	3254	0.298	3275	0.3	0.62%	0.66%
W4	1858	3244	0.294	3327	0.3	2.50%	1.86%
W5	1874	3272	0.293	3272	0.3	0.01%	2.42%
W6	1874	3272	0.304	3263	0.3	0.27%	1.28%
W7	1867	3260	0.292	3274	0.3	0.41%	2.64%
W8	1872	3269	0.292	3341	0.3	2.17%	2.50%
Average	1864	3254	0.294	3323	0.3	2.09%	2.17%

Vp * = P-wave velocity estimated by Equation (8).

In an elastic solid medium, the velocities of P- and R-wave propagation depend on elastic constants and the mass density of the material. The relations among P- and R-wave velocities can be expressed by Equations (9) and (10) [3]. From these relationships, the elastic properties, such as the elasticity and strength, can be estimated. We checked the accuracy of our experimental data by comparing the dynamic moduli from R- and P-wave velocities by the MASW and the ultrasonic test to one determined from the static modulus in Table 4. It shows that errors of the dynamic moduli for R- and P-wave velocities are 4.37% and 8.82%, respectively, and that estimation of the dynamic elasticity by the proposed MASW test is quite accurate.

$$V_P = \sqrt{\frac{E(1-\nu)}{\rho(1+\nu)(1-2\nu)}} \tag{9}$$

$$V_R = \frac{0.87+1.12\nu}{1+\upsilon}\sqrt{\frac{E}{\rho}\frac{1}{2(1+\nu)}} \tag{10}$$

where V_P and V_R are the velocities of P- and R-waves, respectively, and E, ρ and ν are dynamic elasticity, density and Poisson's ratio, respectively.

Table 4. Comparison of dynamic moduli from dynamic and static tests.

Measurement	Dynamic Test		Static Test				Error (%) of Ed
	Avg. (m/s)	Ed * (GPa)	v	Fck (Mpa)	Ec (GPa)	Ed (GPa)	
R-wave by MASW	1864	22.23	0.16	14.84	17.68	21.3	4.37%
P-wave by UPV	3323	23.18	0.16	14.84	17.68	21.3	8.82%

Ed * = dynamic modulus estimated by Equations (9) and (10).

6. Conclusions

The practical application of MASW associated with the MDAFS testing configuration allowed us to create reliable dispersion curves of several fundamental Lamb wave modes. Furthermore, the summed amplitude in the frequency domain equivalent to the IE test results could be extracted from the MASW data. The unique features that make the MASW approach particularly efficient are related to its simple and objective Lamb wave analysis, as well as to its effective data acquisition system. All of the results by the simple Lamb wave analysis associated with the practical MASW test agrees well with data from the traditional UPV test and with the actual thickness values. Thus, this analysis has been verified to be a robust and promising test method for identifying the elastic wave velocities and dimensions. In addition, all of our experimental data are statistically stable and significant. Based on the data presented in this study, we draw the following conclusions.

- The practical MASW approach associated with the multiple sensors and instrumented impact hammer with the appropriate frequency range in characterizing the elastic properties of concrete enables to complete the effective dispersion spectrum especially for the A0 and S1 lamb modes.

- The proposed Lamb wave analysis provides a simpler procedure than the inversion analysis in the general MASW, which has a complicated iterative process that requires changing various material properties such as the density, elasticity, thickness, and Poisson's ratio. Moreover, it showed good accuracy with an error less than 3% when compared to the UPV test results and the actual dimensions.

- The experimental data from MASW are more statistically stable than data obtained from other UPV tests because of improvements in the measurement equipment and analysis; this improvement is associated with the fact that multiple sensors are used.
- We verified that the MASW data are reliable and can be used for the identification of the dimensions, as well as for the estimation of elastic wave velocities for the condition assessment. Our experimental results offer practical guidelines for the application of the MASW method to concrete structures.

Acknowledgments

This work was supported by a National Research Foundation of Korea (NRF) grant funded by the Korean government (MSIP) (NRF-2013R1A2A2A01067754).

Author Contributions

Young Hak Lee and Taekeun Oh conceived and designed the experiments; Young Hak Lee and Taekeun Oh performed the experiments; Young Hak Lee and Taekeun Oh analyzed the data; Young Hak Lee and Taekeun Oh contributed reagents/materials/analysis tools; Young Hak Lee and Taekeun Oh wrote the paper.

Conflicts of Interest

The authors declare no conflict of interest.

References

1. Wu, T.; Fang, J.; Liu, G.; Kuo, M. Determination of elastic constants of a concrete specimen using transient elastic waves. *J. Acoust. Soc. Am.* 1995, *98*, 2142–2148.

2. Boyd, A.; Ferraro, C. Effect of curing and deterioration on stress wave velocities in concrete. *ASCE J. Mater. Civil Eng.* **2005**, *17*, 153–158.

3. Jones, R. *Non-Destructive Testing of Concrete*; Cambridge University Press: London, UK, 1962.

4. Carino, N.J. Stress wave propagation. In *Carin Handbook on Nondestructive Testing of Concrete*; Malhotra, V.M., Ed.; CRC Press: Boca Raton, FL, USA, 2004.

5. Sansalone, V.; Streett, W.B. *Impact-Echo: Non-Destructive Evaluation of Concrete and Masonry*; Bullbrier Press: Jersey Shore, PA, USA, 1997.

6. Udegbunam, O.; Yaman, I.O.; Aklan, H.; Hohm, T. *Developing a Rapid Measure of Concrete Permeability for Use in QNQC Specifications*; Paper No. 990565; Transportation Research Board: Washington, DC, USA, 1999.

7. Kee, S.; Oh, T.; Popovics, J.S.; Arndt, R.; Zhu, J. Nondestructive Bridge Deck Testing with Air-Coupled Impact-Echo and Infrared Thermography. *ASCE J. Bridge Eng.* **2012**, *17*, 928–939.

8. Baggens, O.; Ryden, N. Lamb wave plate parameters from combined impact-echo and surface wave measurement. In Proceedings of the International Symposium NDT in Civil Engineering, Berlin, Germany, 15–17 September 2015.

9. Lamb, H. On waves in an elastic plate. *Proc. R. Soc. A* **1917**, *93*, 114–128.

10. Viktorov, I.A. *Rayleigh and Lamb Waves*; Plenum Press: New York, NY, USA, 1967.

11. Rogers, W.P. Elastic property measurement using Rayleigh-Lamb waves. *Res. Nondestruct. Eval.* **1995**, *6*, 185–208.

12. Laurent, J.; Royer, D.; Prada, C. Temporal behavior of laser induced elastic plate resonances. *Wave Motion* **2014**, *51*, 1011–1020.

13. Gilchrist, M.D. Attenuation of ultrasonic Rayleigh-Lamb waves by small horizontal defects in thin aluminum plates. *Int. J. Mech. Sci.* **1999**, *41*, 581–594.

14. Pei, J.; Degertekin, L.; Khuri-Yakub, B.T.; Saraswat, K. *In situ* thin film measurement with acoustic Lamb waves. *Appl. Phys. Lett.* **1995**, *66*, 2177–2179.

15. Achenbach, J.D. Lamb waves as thickness vibration superimposed on a membrane carrier wave. *J. Acoust. Soc. Am.* **1998**, *103*, 2283–2286.

16. Royer, D.; Dieulesaint, E. *Elastic Waves in Solids I: Free and Guided Propagation*; Springer: Heidelberg, Germany, 2000.

17. Ryden, N.; Park, C.B.; Ulriksen, P.; Miller, R.D. Lamb wave analysis for non-destructive test-ing of concrete plate structures. In Proceedings of the Symposium on the Application of Geophysics to Engineering and Environmental Problems (SAGEEP 2003), San Antonio, TX, USA, 6–10 April 2003.

18. Gibson, A.; Popovics, J.S. Lamb Wave Basis for Impact-Echo Method Analysis. *ASCE J. Eng. Mech.***2005**, *131*, 438–443.

19. Graff, K.E. *Wave Motion in Elastic Solids*; Oxford University Press: London, UK, 1975.

20. Park, C.B.; Miller, R.D.; Xia, J. Multichannel analysis of surface waves. *Kans. Geol. Surv. Geophys.***1999**, *64*, 800–808.

21. Park, C.B.; Miller, R.D.; Xia, J.; Kansas Geological Survey. Imaging dispersion curves of surface waves on multi-channel record. In Proceedings of the 68th Annual International Meeting Society of Exploration Geophysicists, Houston, TX, USA, 25–30 October 1998; pp. 1377–1380.

22. Popovics, J.S. *A Study of Static and Dynamic Modulus of Elasticity of Concrete*; ACI-CRC Final Report; University of Illinois: Urbana, IL, USA, 2008.

23. ACI Committee 39. *Standard Test Method for Compressive Strength of Cylindrical Concrete Specimens*; American Concrete Institute: Farmington Hills, MI, USA, 2014.

24. ACI Committee 469. *Standard Test Method for Static Modulus of Elasticity and Poisson's Ratio of Concrete in Compression*; American Concrete Institute: Farmington Hills, MI, USA, 2014.

25. Lydon, F.D.; Balendran, R.V. Some observations on elastic properties of plain concrete. *Cem. Concr. Res.* **1986**, *16*, 314–324.

26. *British Standards 1881-203, Recommendations for Measurement of Velocity of Ultrasonic Pulses in Concrete*; British Standards Institute: London, UK, 1987.

27. Oh, T. Defect Characterization in Concrete Element Using Vibration Analysis of Contactless Impact-Echo Test Data. Ph.D. Thesis, University of Illinois, Urbana, IL, USA, 2007.

28. ACI Committee 214R. *Guide for Evaluation of Strength Test Results of Concrete*; American Concrete Institute: Farmington Hills, MI, USA, 2011.

© 2016 by the authors; licensee MDPI, Basel, Switzerland. This article is an open access article distributed under the terms and conditions of the Creative Commons Attribution (CC-BY) license (http://creativecommons.org/licenses/by/4.0/).

Index